THE FORBIDDEN UNIVERSE

THE FORBIDDEN UNIVERSE

The Occult Origins of Science and the Search for the Mind of God

Lynn Picknett and
Clive Prince

A Herman Graf Book
Skyhorse Publishing

Skyhorse Publishing books may be purchased in bulk at special discounts for sales promotion, corporate gifts, fund-raising, or educational purposes. Special editions can also be created to specifications. For details, contact the Special Sales Department, Skyhorse Publishing, 307 West 36th Street, 11th Floor, New York, NY 10018 or info@skyhorsepublishing.com.

www.skyhorsepublishing.com

10 9 8 7 6 5 4 3 2 1

Library of Congress Cataloging-in-Publication Data available on file

PEFC

PEFC/16-33-111
CATG-PEFC-052
www.pefc.org

In loving memory of

Lily Iris Prince (1922–2010)

David William Prince (1922–2009)

Unless you make yourself equal to God, you cannot understand God; like is understood by like. Make yourself grow to immeasurable immensity, outleap all body, outstrip all time, become eternity and you will understand God. Having conceived that nothing is impossible to you, consider yourself immortal and able to understand everything, all art, all learning, the temper of every living thing. Go higher than every height and lower than every depth. Collect in yourself all the sensations of what has been made, of fire and water, dry and wet; be everywhere at once, in land, in the sea, in heaven; be not yet born, be in the womb, be young, old, dead, beyond death. And when you have understood all these at once – times, places, things, qualities, quantities – then you can understand God.

'Mind to Hermes' (*Corpus Hermeticum Treatise XI*)
Translated by Brian P. Copenhaver

CONTENTS

INTRODUCTION

In September 2010 the London *Times* carried the banner headline 'Hawking: God did not create Universe', conveying a sense of finality, as if one man – no matter how distinguished – had finally answered arguably the greatest question of all time. In fact, to us the most remarkable thing about this was that Britain's leading broadsheet thought this topic worthy of their front page. Although it was publishing extracts from his latest book, *The Grand Design,* the readiness with which *The Times* accorded Hawking not only the headline, but also a lengthy article and most of the accompanying magazine, shows just how big the debate between religion and science has become.

An even more strident anti-God voice is, of course, that of Richard Dawkins, the British evolutionist and crusading atheist, whose *The God Delusion* (2006) polarized the controversy and gave rise to a flurry of books either attacking him or turning him into a demi-god in his own right. This even led to the bizarre sight of London's big red buses carrying posters that declared, 'There's probably no God. Now stop worrying and enjoy your life', followed swiftly by the other side's call to arms, 'There definitely is a God. So join the Christian Party and enjoy your life'. Seeing these buses sail past in the capital of arguably the most secular country in the West was indeed a curious sight. The controversy has become so cool that it has even found its way into the routines of the edgier comics such as Eddie Izzard and Ricky Gervais, both of whom are vociferously and colourfully atheist.

The debate is by no means simply confined to personal belief or philosophical interest. Religion is now also a hot topic for politicians and social workers, as the gulf widens between the secular and religious mindsets. It seems that virtually every day the media carries some manifestation of this tension, from the French ban on the wearing of the Muslim *burqa* to the fundamentalism that fuels the War on Terror.

When the argument about the existence of God is framed, as it usually is, in terms of dogmatic organized religion, the Dawkins' school seems to be well ahead. When he is arguing with a Christian fundamentalist or a fervent Catholic it is hard not to agree with him. But when he extends his reasoning to anything that touches on the mystical, magical or transcendental, that is where we part company.

There are several major problems with the position advocated by Dawkins and his even more vociferous fellow atheist Christopher Hitchens, author of *God is Not Great* (2007). The first is that, taking advocacy of rationalism and science to its logical conclusion runs the risk of scientism – science as an ideology instead of an objective method for evaluating and improving the natural world. This would create a society in which every aspect of life – not just technology, medicine and so on – is assessed and governed by science. However, as very few people have either the time or the inclination to keep up to date with cutting-edge science, they would have to take the pronouncements of scientists on trust – or faith. Which is exactly how priests rose to power, by claiming an exclusive insight into God's laws beyond the reach of ordinary folk.

We would be back where we started; scientists would be the new priesthood, and scientism would have become the new religion.

More importantly, it seems to us that a sweeping dismissal of anything remotely spiritual or mystical actually ignores a major part of what it is to be human. The Dawkins/Hitchens school fails to distinguish between, on the one hand, the religious *impulse* that is innate to human beings and, on the other, the systems of authority and control that the organized religions have become.

The debate is almost always portrayed with just two alternatives, scientific atheism and organized, dogmatic religion. But something is missing: the profound sense of the 'Other', or the transcendental – what may be termed the mystical, or even magical – that underpins, but is not the same as, religious sensibilities. And, as this book hopes to demonstrate, this is by no means incompatible with a truly scientific worldview.

There has never been a culture – from rainforest tribes to the greatest civilizations such as Rome, ancient Egypt or even the modern West – which did not begin with an understanding of the world based on a belief that it is both purposeful and meaningful, arising from a supernatural ordering of things. It, and everything in it, are here for a reason. This way of looking at the world around us is not learned, but instinctive; it comes naturally to the individual. And this yearning for the transcendental is not rooted in organized religions; they and their priesthoods might exploit this innate impulse, but they did not create it.

Ours is the first civilization where a significant number of people have attempted to break away from such a worldview. But, as Richard Dawkins laments, it is a slow and difficult struggle, precisely because such thinking is second nature to our species. It is so universal, so taken for granted, that it seems to be hardwired into us.

Indeed, while we were writing this book, new evidence emerged, in the work of developmental psychologist Professor Bruce Hood of Bristol University, who concluded at the 2009 meeting of the British Science Association that 'superstition is hardwired', being there from the beginning:

> Our research shows children have a natural, intuitive way of reasoning that leads them to all kinds of supernatural beliefs about how the world works. As they grow up they overlay these beliefs with more rational approaches but the tendency to illogical supernatural beliefs remains as religion.[1]

Hood demonstrated just how hard that wiring is. For example, his study of a group of staunch atheists revealed that even they found the idea of receiving an organ transplant from a murderer utterly abhorrent – a completely irrational reaction. Another researcher, American anthropologist Pascal Boyer, concludes:

> Religious thinking seems to be the path of least resistance for our cognitive systems. By contrast, disbelief is generally the work of deliberate, effortful work against our natural cognitive dispositions.[2]

Hood and Boyer are not talking about deeply mystical and religious feelings but something much more common. Yet, while recognizing how fundamental magical thinking is to human beings, they fail to explain the big question of *why* this should be so.

Similarly, the Dawkins school pays little attention to this mysterious human propensity for a belief in the supernatural and the magical. As it is the very antithesis of scientific, rational thought they don't even give it a nod. But this is dodging a major question. Even if it *is* all just superstition, surely investigating such a basic instinct with an open – a truly scientific – mind would reveal something important about humanity? If, as Dawkins insists, God is a delusion, why should we be *programmed* to be quite so delusional?

As a specialist in the genetic basis of human and animal behaviour, Dawkins has attempted to explain the ubiquity of religion as a by-product of a useful evolutionary trait, suggesting that human beings have evolved an instinct to obey the commands of elders because, as children, we need to do so to remain safe in a dangerous world. We are programmed to believe what we are told by those we look up to in authority. However, as this instinct remains into adulthood we stay susceptible to the pronouncements of authority, and so priests effectively become our surrogate parents, our holy fathers.[3]

Although this makes some sense, it disingenuously addresses only one aspect of religion: why human societies almost always develop religious institutions and priest-hoods – the *exploitation* of magical thinking, not the reason it exists in the first place. Dawkins' scenario would work equally well without religion – if people are programmed to accept authority, then kings and dictators would do just as well, without an appeal to a higher but invisible being.

Science has yet to provide an answer to the basic question of why humans are hardwired to believe. And it is an exquisite irony that one of the products of this magical mindset was science itself. It is, as we will see, what motivated all of the great pioneers of the scientific revolution.

As readers of our previous books will realize, anything that is forbidden has an instant appeal to us. So the discovery that there is a forbidden science was just too tantalizing to ignore. Its focus is an ancient mystical and cosmological system that has always clamoured for our attention, from our first research into Leonardo da Vinci and the Turin Shroud, and our discoveries about the heresy that upholds John the Baptist as the true Christ, which we explore in *The Templar Revelation* (1997) and *The Masks of Christ* (2008). Lynn's *Secret History of Lucifer* (2006), which explores forbidden paths to both mystical and scientific enlightenment, also lit the way to this book.

As we hope to demonstrate, the greatest inspiration of luminaries such as Copernicus and Isaac Newton was almost lost over the centuries. Although the usual explanation for this decline is that scientists simply became too mechanistic – Dawkins would say too sophisticated and intelligent – to think in transcendental terms, we argue that this is not the case, and that there was another reason entirely . . . In fact, this venerable philosophy has much to reveal not only about the origins of science but, we contend, is also increasingly relevant for today's scientists.

This extraordinary tradition is set out in a collection of texts that have had the greatest impact on western culture of any book apart from the Bible, and the greatest impact on the modern world than any book *including* the Bible. Surely that in itself is a major reason for rediscovering these ancient secrets. And the best part is that they are not merely ancient, not just some historical curiosity – they even have something important to teach science of the twenty-first century.

Lynn Picknett
Clive Prince
London, 2010

PART ONE

The Occult Roots of Science

CHAPTER ONE

COPERNICUS AND
THE SECOND GOD

There are three key events that science historians cite as landmarks in the long journey from superstition to intellectual enlightenment: Copernicus' proposal of the heliocentric theory (1543), the prosecution of Galileo by the Church for promoting that theory as fact (1633) and the publication of Isaac Newton's *Principia Mathematica* (1687), which established key physical laws, primarily those of motion and gravity. As a leading historian of science put it: 'The series of developments starting with Copernicus in 1543 and ending with Newton in 1687 maybe be labelled the Scientific Revolution.'[1] However, these great leaps forward were not made because Copernicus, Galileo and Newton elevated pure reason above religious irrationality, but because they were all inspired by the same unashamedly metaphysical and magic-oriented philosophy – one that also excited and motivated other great minds of the time, including our own special hero Leonardo da Vinci.

To today's materialist-rationalists, the unpalatable fact is that a magical mindset not only bubbled along through the Renaissance, but it was magic that inspired and drove the whole of that era's explosion of thought and achievement. In a very real way, magic made the modern world.

The event that is considered *the* watershed moment, the

beginning of the parting of the ways of science and religion, is the proposal of the heliocentric, or 'sun-centred' theory of the cosmos, which posited that the Earth circles around the sun and not, as had been thought, the other way around. The radical new notion was proposed by Nicolaus Copernicus (1473–1543) as the Polish canon Mikolaj Kopernik styled himself in the manner of contemporary scholars.

Until then astronomy and its esoteric twin astrology had traditionally been based on the belief that the Earth was at the centre of the universe. It was a natural assumption, since the sun, moon and stars appear to move around us in regular cycles, while the world we stand on seems to be static. The only complication with this system was presented by the movement of the five planets visible to the naked eye, which despite demonstrating a pattern, did not appear simply to circle the Earth. In the second century CE the Greek-Egyptian astronomer and mathematician Claudius Ptolemaeus, who is known as Ptolemy, devised an Earth-centred model with a complex system of cycles and epicycles to account for the movements of the planets. He was the single great astronomical authority until Copernicus took centre stage.

Strangely, for such a monumentally influential figure, very little is known about Copernicus the man, although the outline of his life is well documented. He was born in Torun in Poland in 1473 to a copper merchant, hence the name. His father died when Copernicus was young, leaving an uncle, who was a canon, to raise him. After studying church law, he extended his stay in the stimulating environment of Renaissance Italy by studying law and medicine at Padua in the Republic of Venice. A gifted artist and draughtsman, his real passion was astronomy, to which he devoted much of his free time.

When his uncle became a bishop, he secured Copernicus a job as a church administrator, or canon, in the town of

Frombork. He lived out the rest of his life, based in a tower – now known as Copernicus' Tower – in the courtyard of the cathedral. His remains were only discovered under the cathedral as recently as 2000. As an ordained clergyman Copernicus was forbidden to marry, but it seems he may not have been totally celibate, according to rumours linking him to his housekeeper. This did not go down well with the Church authorities.

His duties gave him enough leisure time for his passion for astronomy, which he indulged in his tower. Like many astronomers at the time, Copernicus was dissatisfied with the fixes and fudges that were needed to make Ptolemy's system work, and so set out to address the problem. But unlike the vast majority, the results Copernicus achieved would change astronomy for ever.

Copernicus developed his radical new theory in the first decade of the sixteenth century, but refrained from going public for many years, contenting himself instead with scholarly discussions and penning an account for private circulation in the early 1510s. He only published what he termed his 'new and marvellous hypothesis', *On the Revolutions of the Celestial Spheres* (*De revolutionibus orbium coelestium*) at the end of his life – the last page proofs were delivered to him on his deathbed in 1543. The popular science writer Paul Davies calls the book 'perhaps the very birth of science itself'.[2]

Contrary to common belief, Copernicus did not delay publication until death made him safe from the Vatican's wrath. He was only reticent about going public because of the academic controversy his theory would generate, and only agreed to write his book under pressure from colleagues who were excited by his theory. Even Pope Paul III had listened enthusiastically to a lecture on the subject given by his secretary, the German scholar Johann Widmannstetter, ten years before *On the Revolutions* was published. A cardinal who attended the lecture, the Archbishop of Capua, was one of those who urged Copernicus to write and publish his theory. So much for today's perception of the Church's hostility.

On the Revolutions put forward three new controversial ideas: That the Earth moves in space, revolves on its own axis and that it and the other planets circle the sun. Copernicus pointed out flaws in the old Ptolemaic system and set out the observations that led him to propose a new model of the universe. On the thirty-first page he reveals his groundbreaking, even shocking, proposition in the form of a diagram that shows the planets, in their correct order, circling the sun. And just four lines beneath the all-important diagram he makes an extraordinary statement:

> Accordingly [considering the sun's central position], it is not foolish that it has been called the lamp of the universe, or its mind, or its ruler. [It is] Trismegistus' visible God . . . [3]

Copernicus was linking the sun's physical place in the solar system to resolutely transcendental concepts: that the sun is the universe's 'mind' or the seat of the power that rules all creation, or 'Trismegistus' visible God'. And it is in those three words that the greatest clue to understanding Copernicus' theory lies, for they reveal a hint of the *real* heresy that was to rock the Vatican to its foundations.

MAN THE MIRACLE

To discover why Copernicus' reference was – and in certain respects still is – so earth-shaking we have to look back at another seminal document, published over half a century earlier, which cited the same mysterious authority.

Here was a tract that many have called the manifesto of

the Renaissance,[4] as it crystallizes and embodies the spirit and purpose of that new era. Published in Rome in 1487, it has become known as the *Oration on the dignity of man* (*De hominis dignitate*). Intended to be given as a public lecture, but never delivered, it was written by the twenty-four-year-old Giovanni Pico della Mirandola (1463–94). As the youngest son of the ruler of the city-state of Mirandola in northern Italy, and Prince of Concord, Pico's name was already known. Although his family may have been only B-list nobility it was related by marriage to illustrious dynasties such as the Sforzas of Milan and the Estes of Ferrara. Pico had inherited influence, which he was happy to exploit.

When he arrived in Rome from Florence, after attending various universities, including Paris, Pico had with him a set of nine hundred theses – statements from various philosophical, mystical and esoteric traditions – which, he claimed, were mutually consistent and reconcilable. He said he would demonstrate this in a public debate before Rome's intelligentsia. But as the majority of his sources were not Christian, his request for a public debate was refused and his work condemned. This was Rome, after all.

Pico was, however, not to be dismissed so easily. With astonishing courage and foolhardiness (a combination that distinguishes many Renaissance heroes), he published an *Apology* – in fact, a defence – which included his nine hundred theses and what would have been his opening speech in the debate, the *Oration on the Dignity of Man*.

As his chosen title suggests, Pico's fundamental point concerned the brilliance of humankind and its privileged place in creation. To him, a human being's defining faculty is his intellect, the hunger for knowledge and the ability to satisfy it.

According to Pico's parable, after God made the universe and populated it with the angelic beings of heaven and the beasts of the Earth, each with its specific nature and function, he still needed a creature 'to think on the plan of his great work'.[5] As every niche in the cosmological eco-system had already been filled, God decreed that Man should 'have joint possession of whatever nature had been given to any other creature'.[6] Furthermore, being of an 'indeterminate nature' that was 'neither of heavenly nor earthly stuff, neither mortal nor immortal',[7] Man could choose with his own free will the attributes of any other created being, earthly or celestial. Only Man has the flexibility to choose his own path:

> . . . with the sharpness of his senses, the acuity of his reason, and the brilliance of his intelligence [he is] the interpreter of nature, the nodal point between eternity and time.[8]

Aligning humanity with angels was fundamentally anath-ema to the Church of Rome, for whom the doctrine of original sin means that humans are born physically and spiritually soiled, only reaching Heaven if they submit to the Church's dogma and the pronouncements of its priests. And perhaps not even then.

Pico's landmark *Oration* opens with an appeal to two authorities. The first is 'Abdala the Saracen', the ninth-century Muslim scholar Abd-Allah ibn Qutaybah, who declared there was nothing more wonderful in the world than Man. Pico follows with a quotation from the same mysterious sage whom Copernicus would also come to cite: 'the celebrated exclamation of Hermes Trismegistus, "What a great miracle is man, Asclepius" confirms this [Abdala's] opinion'.[9]

It is easy to see why Pico found himself in such hot water. It was not the best idea to start a debate with Holy City scholars by appealing to the authorities of both a Muslim and a resolutely non-Christian sage, Hermes Trismegistus. Notably, his theses also gave pride of place to the Cabala, the Jewish mystical system (which is very different from the modern cult popularized by Madonna).

Pico's *Apology* only made matters worse. Under pressure from Roman scholars, Pope Innocent VIII swiftly banned it. In the interests of self-preservation Pico retracted his claims, before prudently fleeing to Paris. But the Pope's arm was long, and even there he was imprisoned. Yet, as we will see, just when all seemed lost, Pico's fortunes were to turn around.

Pico's *Oration* is illuminating about the Renaissance for several reasons. It reveals the era's defining characteristic, a dramatic shift in attitudes about humanity: Man suddenly became a being of wonder with limitless abilities and possibilities rather than a miserable creature innately blighted and damned by original sin. It also highlights the clash between two mindsets: the new, open, questioning, eclectic spirit of the Renaissance – in particular its willingness to take seriously sources of wisdom outside the Christian domain – and the old, blinkered, Bible-bound attitude of the Middle Ages. The Church had always been wary of learning for learning's sake, frowning on novelty and intellectual challenge. The frenzy of interest in new ways to explore the universe and humankind's place within it was the direct result of being freed from the old shackles. Effectively, the Renaissance represented a great surge of collective self-confidence.

To 'think for oneself' today often implies a rejection of established religion and all forms of 'superstition', however this was emphatically not the case among the intellectuals of Renaissance Europe. Most of the traditions from which Pico drew his theses were not established works of physics or mathematics, but metaphysical, mystical and what we have come to know as occult sources. Above all it was the works of Hermes Trismegistus that drove Pico with a passion.

There were many reasons why the Renaissance happened when it did. One was the renewed interest in the scholars and philosophers of ancient Greece and Rome, especially Plato. Many works from antiquity had been lost to Europe but preserved in the Middle East, from where they began trickling back during the late Middle Ages. This became a flood in 1453 when Constantinople, the last bastion of the Byzantine Empire (itself the last bastion of the Roman Empire), fell to the Muslim Ottomans. Another factor was the expulsion of the Jews from Spain in 1492, their scholars dispersing into Europe's intellectual centres. Jewish traditions of learning had until then been ignored in Christian Europe.

Apart from the intellectual sphere, cultural, economic and political factors all played a part in giving birth to the Renaissance. The fact that its first flowering took place in Florence, for example, was intimately linked to the city's wealth as well as its republican government.

One of the most important defining factors of the Renaissance, however, was a renewal of interest in the esoteric, specifically the theory and practice of magic. Given the scale of its impact on the Renaissance and the fact that it was hardly hidden away (as Pico's *Oration* clearly demonstrates), it is astonishing that historians completely ignored the influence of this renewed interest on the period until the 1940s, when studies began to reveal its influence over many of its great figures. It is only really over the last half-century or so that the crucial importance of esoteric, magical philosophies has been properly appreciated, as for example in the work of academics such as British historian Frances A. Yates (1899–1981). In a series of books published in the 1960s and 70s, Yates demonstrated that the Renaissance was largely motivated and driven by the 'occult philosophy', a blend of fifteenth- and sixteenth-century magical and esoteric systems.

The term 'occult philosophy' comes from one of the period's most important expositions of the principles of magic, *Three Books on the Occult Philosophy* (*De occulta philosophia libri tres*) by Heinrich Cornelius Agrippa, published 1531–33. The Latin *occultus* simply meant hidden, obscured or, by extension, secret, but not necessarily supernatural. Agrippa's book would have been understood at the time it was published to be about 'hidden philosophy'.

Magic's reputation enjoyed a major boost in the Renaissance. From being the exclusive province of reclusive, more than usually malodorous and scary individuals, it very nearly became mainstream, and was widely discussed as a respectable aspect of philosophy and even theology. In his *Oration*, for example, Pico della Mirandola argues that magic is a valid path to knowledge, but is careful to differentiate between the more odius and hellish magic that utilises demons, and the natural magic that comprises the highest realisation of philosophy.[10] In the intellectual explosion that was the Renaissance, magic came to be considered an integral part of all aspects of human knowledge.

As Frances Yates demonstrated, the Renaissance occult philosophy was based on three streams of esoteric thought. Of the three, modern academics favour what is now known as Neoplatonism, a philosophy and cosmology developed in the intellectual melting-pot of the Egyptian seaport Alexandria in the second and third centuries CE. Neoplatonism blended the original ideas – then already eight hundred years old – of the great Greek philosopher Plato with other Greek and Egyptian mystical concepts. A second strand was a Christianized version of the Jewish Cabala, which Pico aligned with the occult philosophy in what was to be considered his greatest innovation. But the third, and by far the most important strand, was Hermeticism,[11] the philosophy attributed to the legendary wise man honoured by Pico and Copernicus: Hermes Trismegistus, or the 'Thrice-Great Hermes'. And it is this strand that shifted the world from a morass of ignorance and self-hate to the sunlit uplands of intellectual genius.

The sheer power of Hermeticism cannot be over-estimated. It effectively created the Renaissance, whose essence could be summed up by Hermes' adage, '*Magnum miraculum est homo*' (literally, 'Man is a great miracle'). Hermeticism embraced that fanatical determination to discover, invent and understand, and the overwhelming sense of excitement at the prospect of endless possibilities. It seized the imaginations not only of Copernicus but also later luminaries. It drove them, hearts and minds, to dare to challenge the old thinking and encompass the most radical, even subversive, ideas – which changed the world forever. Their contributions to science would simply have been impossible without Hermeticism. Without Hermes Trismegistus, these great thinkers would never have fully realized their genius.

KEEPER OF ALL KNOWLEDGE
Hermes Trismegistus was a legendary Egyptian sage and teacher, whose wisdom was embodied in a collection of books known as the Hermetica. Although during the Renaissance Hermes Trismegistus was taken to be his full name – hence Copernicus simply calling him 'Trismegistus' – 'Thrice-Great' is an honorific, so his proper name is just 'Hermes'. He was said to be a descendant of the god Hermes, or his Roman equivalent, Mercury.

During the Middle Ages, Hermes Trismegistus was a truly legendary figure, known only from odd fragments of his own supposed writings and references to him and his work in ancient texts. One such reference came from

Clement, Bishop of Alexandria, who around 200 CE witnessed Egyptian priests and priestesses parading their sacred books and noted that there were forty-two works of Hermes. (Which, if nothing else, according to cult comedy science-fiction writer Douglas Adams, is a number that is sacred to galactic hitch-hikers.)

Although scattered references to the Hermetica survived, all but one of the actual books had disappeared, at least in Europe. However, hand-written copies of many of the books did still circulate in Byzantium and, significantly, in Islamic centres of learning. At some point eighteen treatises were grouped together and became known as the *Corpus Hermeticum*. When, by whom and why they were selected, is unknown, but the *Corpus* was finalized by the eleventh century, and Byzantium seems to be the logical location for its compilation.

Another important source on Hermeticism was an anthology of around forty fragments, some from the *Corpus Hermeticum* but others otherwise unknown, compiled by the pagan Macedonian scholar Stobaeus around 500 CE, and including a complete treatise, *The Virgin of the World* (*Korè Kosmou*). Another Hermetic text may only be a mere half page long; the *Emerald Tablet,* but it is difficult to overstate its importance. Allegedly containing the words of Hermes Trismegistus himself, the thirteen alchemical maxims of the *Emerald Tablet* were believed to have originally been engraved on a tablet fashioned from the bright green jewel itself. Nobody knows for sure if this work has any connection with the Greek Hermetica, since it comes from an Arabic source that entered Europe via Spain in the twelfth century, but it was immensely influential among alchemists, helping cement Hermes' status as more than merely a wise man. To those whose admiration bordered on worship, he was at the very least a semi-divine teacher.

19

The one complete Hermetic book known in Europe in the Middle Ages was the *Asclepius,* or *The Perfect Word,* a fourth-century Latin translation of a lost Greek original, a question-and-answer session between Hermes and his eponymous pupil. Asclepius was the Greek god of healing; the pupil in the treatise is his descendant, although he himself is not divine. The names of the characters, including Ammon and Tat (Thoth) who also appear as witnesses to the debate, reveal the Hermetic attitude to both divinity and humankind in general. This has it that while there is a God, human beings who attain a certain level of wisdom can themselves become divine. An example of this is presented in the form of Asclepius' ancestor, originally a mortal who discovered medicine, and who despite being dead and buried – his mummified body lay in a specially constructed temple – was still able to intercede for the sick. Similarly, Hermes Trismegistus describes himself as a descendant of the god Hermes, who continues to help mankind.

The Hermetic texts are a mixture of on the one hand philosophical and cosmological teaching, and on the other astrology, alchemy and magic. Over the centuries, and even today, attempts have been made to separate the two, on the grounds that the philosophy itself is sophisticated and coherent, while the astrology and magic is considered primitive and incoherent. (One 1920s edition simply deleted this material.) Some even consider the compilation of the *Corpus Hermeticum* as an attempt to purge the canon of the most magically inclined texts. Of all the known Hermetica, those in the *Corpus* are conspicuously the least magical, but even they include some arcane elements -- which is hardly surprising given that the philosophy and cosmology are indivisible from an occult worldview.

THE MIND OF GOD

The Hermetic books explore an intimately related cosmology, philosophy and theology that is fairly accessible in principle, even if some of the details are as abstruse as an ancient alchemical text, and for similar reasons. While any student might read the books, they are designed to speak only to the heart and mind of those who are worthy of learning their secrets. An ability to navigate the extraordinary allusions and metaphors, and an understanding of the connections between them, is in itself a sort of initiation into a world of spiritual and intellectual wonders.

Despite the medieval and Renaissance tendency to regard the books as the work of the great Hermes Trismegistus, they are obviously authored by various individuals who 'present different interpretations of their common doctrine,'[12] and with scrupulous honesty often point out that some of the treatises are contradictory.[13] The reason for the attribution to Hermes is that all of the authors have chosen to remain anonymous, which – as we will see – is very telling. The writers believe that the common doctrine stems from Hermes, God's chosen teacher of humankind, 'the all-knowing revealer'.[14]

The Hermetica's philosophy and cosmology is not only mystical but emphatically magical, embracing different realms of being, from gross matter to the divine spheres, and that of supernatural beings, divine, angelic and demonic. But ultimately it is monotheistic, ascribing all creation to a single God, while also encompassing lesser gods and goddesses, a category to which even mortal humans can aspire if they become sufficiently advanced. 'Advanced' is not merely the sort of 'spiritual evolution' that is today assumed as a badge of superiority by New Agers; great intellectual progress that benefits humanity also qualifies. Asclepius won his godhood for his medical advances. (It certainly beats a Nobel Prize.)

Unlike the creator-God of Judeo-Christian tradition, however, the Hermetic God is intimately part of his creation. In the Hermetic vision, the universe is God and God is the universe. The cosmos is a living entity, and everything in it is imbued with life. Hermeticism also incorporates the once-common idea of the *anima mundi*, the world-soul. The Hermetic universe is really more of a great thought, an emanation of the mind of God, than something zapped into being on his orders. But God needs the universe in order to realize himself, as American historian of science and philosophy Ernest Lee Tuveson writes (his emphasis):

> The essential elements of the Hermetist conception of reality is that the world emanates from the divine intelligence, and, as a *whole* in which *each* part is an essential component member, expresses that great Mind.[15]

As the American philosopher Glenn Alexander Magee – whose speciality is the influence of esoteric thinking, and particularly Hermeticism, on western culture – points out, this explanation of God's need to create the universe overcomes some of the nonsensical aspects of the biblical creation tale. Magee points out that the traditional Judeo-Christian account provides no good reason why God should have wanted or needed to create either the universe or humankind: what does he get out of it? This was one of the main reasons the Hermetic explanation appealed to the increasingly sophisticated Renaissance thinkers: 'The great advantage of the Hermetic conception is that it tells us *why* the cosmos and the human desire to know God exist in the first place.'[16]

Hermeticists see human beings as enjoying a special place in creation. As essentially divine beings stuck in animal bodies, according to Hermeticists, human beings possess not only the divine spark (which is present in everything) but effectively share in God's mind. Humans are the only beings in God's creation with the potential to become divine. Salvation, in the Hermetic scheme, comes from the use of our advanced mystical and intellectual faculties. As Treatise X of the *Corpus Hermeticum* states:

> For the human is a godlike living thing, not comparable to the other living things of the earth but to those in heaven above, who are called gods. Or better – if one dare tell the truth – the one who is really human is above these gods as well, or at least they are wholly equal in power to one another.[17]

One therefore ascends through knowledge, which comes through both greater intellectual and philosophical under-standing of the cosmos and the more spiritual form of enlightenment called *gnosis*. But the relationship between creator and humanity is an endless cycle, as Magee notes:

> Hermeticists not only hold that God requires creation, they make a specific creature, man, play a crucial role in God's self-actualization. Hermeticism holds that man can know God, and that man's knowledge of God is necessary for God's own completion.[18]

So, not only did the Hermetic vision provide a more satisfactory explanation of why the universe exists, it also gave human beings *potentially* the most exalted role – though one that has to be earned. As *Asclepius* declares, 'a human being is a great wonder, a living thing to be worshipped and honoured'.[19] The Hermetica encourages people to use all their faculties, powers and talents in the pursuit of both self knowledge and knowledge of the universe. A major part of the kinship with creation involves observing the world around us and delving deeply to discover its hidden workings. In Hermeticism, this is not mere lofty sentiment, but one of the major paths to salvation. The Hermetic motto 'Follow nature'[20] – which would come to have a profound effect on the beginnings of science – bears witness to this cornerstone of the philosophy.

MAGIC AND MYSTERY AT HARRAN

Wherever and whenever Hermeticism originated, it was being discussed by both Christian and non-Christian writers in the Roman Empire from the second century onwards. But it disappeared soon after Christianity became the dominant Roman religion and persecutor of pagans in the fourth century. Apart from a fragmentary presence, the Hermetica basically vanished from Europe until the Renaissance. But its wisdom survived outside the Christian world, focusing on the city of Harran, some fifty miles south of Edessa in south-eastern Turkey. How it came to be established there is unknown, but presumably Hermeticists fleeing from Christian persecution would provide an answer.

By the time Harran fell into Arab hands in the mid-seventh century it was a renowned centre of learning. Two centuries later, according to tradition – which may or may not be apocryphal – the inhabitants were given a stark choice by the caliph al-Mamun: convert to Islam, be massacred, or identify themselves as one of the 'peoples of the book'. The Qur'an requires tolerance and protection for the latter – such as Jews and Christians – provided they venerate a prophet recognized by Islam.

Unsurprisingly rejecting the option to be massacred, the residents of Harran identified themselves as Sabians, one of the 'peoples of the book' mentioned in the Qur'an.[21] But the Sabian prophet was found in neither the Old nor the New

Testament. Instead they proudly declared him to be Hermes and their holy book the *Corpus Hermeticum*. Fortunately the Qur'an identifies Hermes with the prophet Idris, the Muslim rendering of the Old Testament Enoch. The Sabians of Harran also venerated Asclepius as a prophet and Agathodaimon ('Good Spirit'), a character in the Hermetic dialogues, as a great teacher and an intermediary with God.[22] They went on pilgrimages to the two great pyramids at Giza, revering them as the tombs of Hermes and Agathodaimon.[23]

Soon after the al-Mamun episode was supposed to have happened, the great library of Baghdad, the House of Wisdom (*Bayt al-Hikma*) – which was also a centre for research, translations of foreign works and an observatory – was re-established. Many Sabians moved there, the most eminent of which was the renowned polymath Thabit ibn Qurrah (835–901). It was here, in Baghdad, that the Hermetic books were translated into Arabic. The foundation of Arab science in the Middle Ages was therefore laid by the Sabians, and inspired by the Hermetica.[24]

The Sabians disappeared from Baghdad and Harran during a clampdown on non-Muslims in the middle of the eleventh century. It is possible they became devotees of Sufism, the mystical form of Islam, which aims at individual communion with the divine. Although Sufism had been around for centuries, it underwent a formalization during the eleventh century that was, some think, due to a Sabian influx.[25]

Many specialists have noted that the revival of interest in the Hermetica in Byzantium coincided with the end of Sabian Hermeticism.[26] But was this purely coincidence? Psellus, the Byzantine Platonic philosopher, became the first westerner to write about the Hermetica in half a millennium and many have speculated that Sabians, fleeing persecution, had carried their precious literature with them to Constantinople.

THE REDISCOVERY

One of the great patrons of the early Renaissance was Cosimo de' Medici, scion of the banking dynasty that pretty much owned the republic of Florence. Cosimo was also hugely ambitious in his vision of what he and his court could accomplish, sending agents out in search of key books, and employing one of the great scholars of the age, Marsilio Ficino (1433–99), on massive learned projects and as tutor to his grandson Lorenzo. Cosimo's aim was nothing less than to re-establish Plato's Academy, this time in Florence, with Ficino as its head. The lynchpin of this somewhat ambitious task was the first ever translation of Plato's complete works from Greek into Latin, then the *lingua franca* of scholarly Europe.

Just as Ficino was about to dip his quill into the ink and get started on Plato, an even more exciting prospect presented itself. One of Cosimo's agents, a monk named Leonardo de Pistoia, returned from Macedonia with a Greek manuscript of the first fourteen treatises of the *Corpus Hermeticum*. Ficino records that in 1463 Cosimo ordered him to drop his translation of Plato forthwith in order to concentrate exclusively on the *Corpus Hermeticum* – an urgency that was probably the result of Cosimo becoming gravely ill, and desperately wanting to read the Hermetic books before he died. He got his wish, with a year to spare.

Because of the mysterious aura surrounding Trismegistus and his lost books, this was by far Ficino's most popular work, as evidenced by the many copies of the manuscript and several editions of the first printing of 1471. The discoveries that Ficino's translation made possible sent seismic shockwaves throughout the academic community in Florence and beyond, being widely and feverishly discussed and debated. The books enticed Pico della Mirandola to Florence, where he studied under Ficino between 1484 and 1486, when he departed for Rome with his nine hundred theses. As Tuveson writes in *The Avatars of Thrice Great Hermes* (1982), 'with the translation by Ficino of the *Hermetica* in the fifteenth century, a kind of "new force" had entered the Western world.'[27]

One reason for the excitement generated by the rediscovery of the Hermetica was precisely because it was so radically different from Christianity's stifling view of creation and humanity's place within it. Another was the idea that an ancient original religion, now lost, lay behind all other religions. This was variously known as the *prisca theologia* ('ancient theology'), *prisca philosophia* ('ancient philosophy') or *philosophia perennis* ('perennial philosophy'). Many believed that this ancient, lost religion could be found in Egypt, as even the Bible acknowledged that its civilisation and religion pre-dated that of the Israelites. Indeed, there was even a suggestion that Moses himself learned great secrets from the Egyptians. Given that Hermes Trismegistus was thought to be the renowned sage of ancient Egypt, it was logical that the Hermetica could contain the ancient theology.

Ficino was hugely influential in his own right. His close relationship with these books lured him ever deeper into the Hermetic world, and he began to discern strangely recurring themes. A modern writer on Italian history, Tim Parks, describes Ficino's momentous declaration:

> The whole world, it seemed, had always followed a single faith whose ancient priests included Zoroaster, Hermes Trismegistus, Orpheus, Pythagoras, Plato, St Paul, St Augustine.[28]

Thus, according to Ficino, a secret line of priests linked the ancient pagan and Christian beliefs. Ficino threw himself into trying to recover and reconstruct this 'single faith', concluding that it was a magical current flowing under and linking many otherwise apparently irreconcilable belief systems. From this he developed the idea of 'natural magic', one that worked with the forces of nature rather than by the conjuration of demons or spirits.

The robust joy in life that marked the Hermetic path extended well beyond that of academic study. As American researcher Peter Tompkins writes:

> Ficino regarded sexual desire as a current of energy responsible for the cohesion of the entire universe . . . Ficino even went so far as to recommend the pagan revels of Bacchus (or Pan) as a way of escaping from normal human limitations into an ecstasy in which the soul was miraculously transformed into the beloved god himself.[29]

Ficino's masterwork was *Three Books on Life* (*De vita libri tres*), published in 1489, which was extremely influential on arcane philosophers such as Agrippa. But once again, despite being a synthesis of several magical and philosophical systems, Hermeticism stood firmly as the heart and soul of Ficino's work.

The next step would be from Florence to Rome. Astounding though it may seem to us today, many in the highest echelons of the Catholic Church were sympathetic to the message of Hermeticism, and considered it to be compatible with Christianity.

The Hermetica proclaimed that the material universe was created by a lesser god, or Demiurge, who had been assigned the task by the God of all. In *Asclepius*, God is said to love this second god as 'His own Son',[30] which has obvious parallels with Jesus. In *Pimander*, the first treatise of the *Corpus Hermeticum*, God's creative Word is also described as the 'Son of God'[31] – to some a clear echo of the majestic opening of the Gospel of John: 'In the beginning was the Word'.

Such references led some early Christian proselytes, such as the late-third/early fourth-century author Lactantius, to accept Hermes Trismegistus as a pagan prophet who foresaw the coming of Christ. This view was by no means unanimous: others such as St Augustine ascribed Hermes' foreknowledge to warnings from worried demons. But when the Hermetica was rediscovered in the fifteenth century, at least enthusiasts could argue their case by invoking early Church authorities.

Some thinkers tried hard to find a compromise, accepting the philosophy and cosmology but rejecting the magic, while others, such as Pico della Mirandola, pointed out that the two sides of the Hermetica were inseparable and argued this demonstrated that magic – provided there was no occult nastiness such as conjuring spirits – was a legitimate Christian activity. After all, Moses had engaged in magical contests with the pharaoh's magi and had probably learned magic in Egypt. Some even suggested that Jesus had performed his miracles by means of natural magic.

Others went further, seeing Egypt as the origin of the wisdom inherited first by the Jews and then by the Christians. This, they argued, elevated Hermes to at least an equal footing with Moses, who despite not being a Christian, was still deserving of respect for his contribution to the religious tradition into which God had chosen to send his son.

The extent to which men in high places accepted this reasoning – even, astonishingly, including the Pope himself – can be demonstrated by resuming the story of Giovanni Pico della Mirandola, who we left earlier languishing in a Parisian prison after his arrest on the orders of Pope Innocent VIII. He didn't languish for long. As Pico was from a well-connected family, his powerful supporters interceded with the Pope on his behalf. One such supporter was Charles VIII of France, and another Lorenzo de'

Medici – 'the Magnificent' – who was now one of the wealthiest and most powerful men in Florence. Eventually the Pope allowed Pico to return to Florence, under Lorenzo's guarantee that he would behave himself, although his works remained on the banned list.

In 1492 Innocent VIII died and was succeeded by the Spaniard Rodrigo Borgia, who wore the papal crown as Alexander VI. His reign certainly began with a bang. He not only absolved Pico and his works from all taint of heresy but wrote him a fan letter, and the fact he did so early in his reign demonstrates how strongly he felt about it. Tantamount to a papal endorsement, the letter was included in subsequent editions of Pico's books. In the event, Pico's repatriation was short lived, as he died in 1494, at the age of just thirty-one.

But why did Alexander support this heretical upstart? As his fan letter suggests, Pico and the Pope shared a passion for all things Hermetic. The Borgia Pope even commissioned tell-tale decorations for his personal rooms in the Vatican – the Appartamento Borgia – which survive to this day. In the series of frescoes on mythological themes by Pinturicchio, Hermes Trismegistus is depicted twice, possibly three times if an image of Mercury slaying the giant Argus is intended as a veiled reference to him.

The first Hermetic reference in Alexander's apartment is in a series of pictures showing the pagan and Jewish prophets who allegedly foresaw Christ's coming. So far this is conventional: images or statues of Hermes Trismegistus appear in several cathedrals for the same reason. More unexpected is a painting in which Hermes and Moses are shown sitting before Isis, implying that Alexander accepted Hermes' equal status to Moses and that both drew their wisdom from Egypt. Judaism is seen as having emerged from the Egyptian Hermetic religion just as Christianity was to emerge from Judaism. Not only does Isis therefore appear in the Vatican, but she is depicted in all her power and glory – not as some pagan deity wretchedly grovelling to a triumphant Christianity.

Other peculiar pro-Egyptian imagery in the Borgia apartments relates to bulls. As that animal was the Borgia family's symbol, this may not be so surprising, at least at first glance. However, the bas-reliefs in Alexander's apartments clearly associate the Borgia bull with the sacred Apis bull of Egypt, which is shown being worshipped and, in turn, worshipping the cross. Once again an association between Christianity and the religion of Egypt is implied, linked thematically with a Borgia pope worshipping Christ, suggesting that the relationship between Hermeticism and Christianity was important to Alexander.

However, extraordinary though it may seem, this is not to imply that Alexander wasn't a Christian, or that a closet occultist had infiltrated the highest office of the Church. It was quite permissible to see Christianity as the heir of a tradition that stretched back to ancient Egypt, and one to be celebrated. Such associations belonged to the new spirit of the time. Indeed, the most surprising thing about the Appartamento Borgia frescoes is that they indicate that even a Borgia pope was capable of caring more deeply about his religion's origins than most Catholics at the time.

THE TRIUMPH OF HERMES

Eighty years after the rediscovery of the lost books of Hermes, Copernicus gave pride of place to the legendary Egyptian sage in his own seminal work on the movements of the planets. But why?

It is hardly surprising that Copernicus was familiar with the Hermetica, having studied in Rome and Padua in the 1480s and 90s, where it was on everyone's lips. But evidence suggests that the works meant considerably more to him than mere intellectual fashion. The debt Copernicus owed to the Hermetica is demonstrated by the fact that the three revolutionary ideas he was to famously propose – the Earth's motion in space, its rotation on its own axis and the orbiting of the Earth and other planets around the sun – *all appeared in the Hermetica.*

Asclepius, for example, provides the following statement in the middle of a discourse on 'classes', or archetypes:

> The class persists, begetting copies of itself as often, as many and as diverse as the rotation of the world has moments. As it rotates the world changes, but the class neither changes nor rotates.[32]

Hermeticism lays great emphasis on the sun, which is regarded as a kind of relay station for God's creative and sustaining power and described in turn as the 'visible god' and a 'second god'.[33] But although it isn't so surprising to find the sun given such prominence in the Hermetica, some passages about its importance are intriguingly specific. Treatise XVI, in which Asclepius expounds various points of teaching to King Ammon, contains two particularly tantalizing statements: 'For the sun is situated at the centre of the cosmos, wearing it like a crown'[34]; and 'Around the sun are the eight spheres that depend from it: the sphere of the fixed stars, the six of the planets, and the one that surrounds the earth.'[35]

These 'spheres' correspond to the modern concept of orbits, as it was thought that the celestial bodies were fixed to transparent spheres. Under the old Ptolemaic system the spheres surround ('depend from') the Earth, with the sun occupying its own sphere. But this is not what is described in Treatise XVI, with the spheres surrounding the sun, which is situated at the centre. And the Earth has its own sphere which, like the other planets, 'depends from' the sun in a way that only makes sense in Copernican terms.

Perhaps most interesting of all is the fact the heliocentric aspects are only mentioned in passing, when some other principle is being elucidated. It appears that the writers of at least these particular Hermetic treatises took the Earth's journey around the sun for granted. Clearly, by referring to Hermes Trismegistus in his own exposition of the helio-centric system – besides quoting from Ficino on the sun as the embodiment of God – Copernicus shows that he was at least familiar with the prototype for his own ideas. As Frances Yates concluded:

> One can say, either that the intense emphasis on the sun in this new worldview was the emotional driving force which induced Copernicus to undertake his mathematical calculations on the hypothesis that the sun is indeed at the centre of the planetary system; or that he wished to make his discovery acceptable by presenting it within the framework of this new attitude. Perhaps both explanations would be true, or some of each.
>
> At any rate, Copernicus' discovery came out with the blessing of Hermes Trismegistus upon its head, with a quotation from that famous work in which Hermes describes the sun-worship of the Egyptians in their magical religion.[36]

While Tobias Churton, the British authority on Hermeticism and Gnosticism, states that (his emphasis):

> One gets the impression that Copernicus is saying: *the truth of the matter was already there, but went unseen because we judged things from an earthly perspective. But Hermes, at the beginning of science, he saw it.*[37]

The fact that Copernicus was inspired by the Hermetica

also, of course, made the debate over heliocentricity of keen interest to Hermeticists, especially as it seemed to vindicate their semi-sacred texts. If the theory could be proven beyond doubt, it would engender confidence in the entirety of the Hermetic philosophy. And as we shall see, there were some who took it considerably further than that. Certainly, and unsurprisingly, in the ensuing furore about Copernicus' new theory, the Hermeticists were among his most ardent supporters.

'TOO MUCH IN THE SUN'

As already mentioned, it is a misconception that the heliocentric theory in itself sparked off a notorious religious furore. Although Copernicus dedicated his book to Pope Paul III, he was not, as many assume, simply boot-licking in an attempt to head off papal disapproval. After all, Paul was quite happy with Copernicus' theories ten years before *On the Revolutions* was published. In the dedication, somewhat airily, Copernicus explained his reluctance to go public by saying he wanted to avoid harsh words from lesser scholars: he was not concerned it might stir up theological controversy, let alone accusations of heresy.

Even the notorious preface, apologetically explaining that the ideas contained therein were just theories, no more valid than any other about the workings of the heavens, was designed to placate *scholars*. The preface was actually written by a Lutheran theologian, Andreas Osiander, who oversaw the printing of *On the Revolutions* after Copernicus' death. But because Osiander didn't make his authorship clear, many readers assumed the preface expressed Copernicus' own position. Georg Rheticus, the mathematician who persuaded Copernicus to go public with his theory, later threatened to beat Osiander up for his audacity.

The heliocentric theory raised no major theological

difficulties anyway. True, there are a handful of implications in the Old Testament concerning the immobility of
the world. The First Book of Chronicles, for example, states
that, 'The world is firmly established; it cannot be moved',[38]
and Joshua is said to have convinced God to stop the sun in
the sky, which implies that it was the sun, not the Earth,
which moves.[39] But in the end few churchmen thought
Copernicus' theory was worthy of oiling the rack and
heating the pincers.

Ironically, any religious objections came not from the
Vatican but from Protestants, although even the most
hellfire-and-damnation regarded the theory as mere folly
as opposed to blasphemy. Martin Luther himself ridiculed
it, but mainly because he was aghast at the suggestion that
astronomy could have got it so fundamentally wrong for
so long.

This was also largely the position of scholars, who too
were disturbed for another reason, which is less obvious
today. Proposing that traditional astronomy was profoundly flawed seemed intimidating, since it implied that
human understanding of the order of the universe, and the
way one part influenced another, was seriously lacking. If
Copernicus was right, then *everything* changed.

This was not yet the era of science as we know it in the
modern sense. Even learned men such as Copernicus and
Johannes Kepler believed that a greater understanding of
the movements of the heavenly bodies would improve the
accuracy not only of astronomy but also its esoteric twin,
astrology. No astronomer at that time believed the
workings of the universe were due to impersonal physical
forces. To them, God had decreed that the universe should
operate in the way it did. As such, discovering how it
worked offered an insight into the divine mind, and might
also throw light on God's plan for all creation. This mindset
drove the likes of Kepler who, building on Copernicus'

work, established the laws of planetary motion.

Kepler (1571–1630) was another great name of the scientific revolution who was steeped in the Renaissance occult tradition. He believed that the planets, including the Earth, are living entities with their own world souls and that the seat of the *anima mundi* is in the sun. As an astrologer he wrote that a new star that appeared in 1604 portended major changes on Earth. Unsurprisingly, his writings also reveal a detailed knowledge of the *Corpus Hermeticum*.

A suggestion that Kepler drew direct inspiration from the works of Hermes Trismegistus appears in the following enigmatic statement from the *Harmony of the World* (*Harmonices mundi*), in which he outlined the laws of planetary motion:

> . . . after the pure Sun of that most wonderful study began to shine, nothing restrains me; it is my pleasure to yield to the inspired frenzy, it is my pleasure to taunt mortal men with the candid acknowledgement that I am stealing the golden vessels of the Egyptians to build a tabernacle to my God from them, far, far away from the boundaries of Egypt . . . See, I cast the die, and I write the book.[40]

Some embraced Copernicus' new ordering of the solar system as a leap forward in understanding the workings of creation, but it absolutely terrified many others. If the traditional understanding of cosmological behaviour was wrong, then how could men begin to understand their own place in the universe? And the uncertainty – some accepted Copernicus' new order, others stuck to the old system of Ptolemy – meant that chaos reigned, and not merely in the academic discipline of astronomy, but in the world at large. This aspect of the heliocentric debate was so significant at the time it even surfaces as a major theme in William

Shakespeare's *Hamlet*. Shakespeare was obviously familiar with Hermeticism, as allusions appear in his works, for example in Hamlet's homage to humankind which echoes Pico's vision: 'What a piece of work is a man! How noble in reason! How infinite in faculty! . . . In action how like an angel! In apprehension how like a god!'[41]

Astronomers, rather than literary historians, have often seen clear and specific allusions to the debate over the heliocentric theory in the play, which dates from around 1600. Peter D. Usher, Professor Emeritus in Astronomy and Astrophysics at Penn State University, has recently argued that the whole work is an allegory for the struggle between the two models of the universe, suggesting that *the* major theme is that Hamlet, prince of the new learning and repeatedly associated with the sun, is involved in a bid to establish his rightful place as the king – at the centre of his universe – by overthrowing his uncle Claudius. It just so happens that Ptolemy's first name was Claudius.

References to the heliocentricity controversy are undeniably scattered throughout the play. For example, Hamlet writes to his love interest Ophelia:

> Doubt that the stars are fire;
> Doubt that the sun doth move;
> Doubt truth to be a liar;
> But never doubt I love.[42]

Other references are less obvious today. For example, many generations of readers and actors have studied Hamlet's apparently peculiar declaration, 'I could be bounded in a nutshell and count myself a king of infinite space',[43] without realizing its potentially subversive undercurrent.

The leading supporter of Copernicus' theories in Shakespeare's England was the mathematician (and Member of Parliament) Thomas Digges, who went one step further than his hero. Although Copernicus maintained the traditional belief that the stars all exist on the same sphere, equally distant from the centre of the solar system, Digges suggested that they are positioned at different distances in an infinite universe. His actual words were that the world was not enclosed in the stellar sphere *'as in a nutshell'*. And as Shakespeare knew Digges personally – they lived in the same building in Bishopsgate, east London, and Digges' son worked at the Globe Theatre[44] – there seems little doubt the 'nutshell' line was an allusion to Digges' theory.[45]

But the most specific of Shakespeare's references to the heliocentric debate relate to Tycho Brahe (1546–1601), the flamboyant Danish alchemist and astronomer (whose eccentric household included a clairvoyant dwarf who lived under his table and a pet elk that met its end in a drunken plunge down stairs). Tycho's great ambition was to reconcile 'the mathematical absurdity of Ptolemy and the physical absurdity of Copernicus'[46] through a hybrid model in which the sun and moon orbit the Earth but the other planets and stars orbit the sun. Tycho therefore literally embodied the struggle between the two great systems.

Tycho was employed by his patron, Frederick II of Denmark, to purchase artworks and scientific equipment for his new castle at Elsinore (built just twenty-five years before *Hamlet* was written), where the play is set. Frederick gave Tycho the island of Hven, in sight of the castle, to build an observatory, Uraniborg. The character of Hamlet, like Tycho, was a graduate of the University of Wittenberg. Most tellingly, two of Tycho's relatives were envoys to London in Shakespeare's day. Their names – Frederick Rosenkrantz and Knud Gyldenstierne – are the same as Hamlet's ill-fated peers, Rosencrantz and Guildenstern.

Obvious though the links may be, what was Shakespeare trying to convey about the big heliocentric debate? After all, the play sees the demise of all of its leading characters, including Hamlet himself, in the famously bloody finale. So although Shakespeare seems to be championing the new Copernican system, his major emphasis is really the uncertainty that was overturning the world and throwing everything into chaos.

During Shakespeare's time, none of this was an issue for the Church, which had long frowned on astrology. But by Galileo's day heliocentricity had become a burning issue and its spokesmen were condemned as heretics. He was first warned off in 1616, and it was only in that year – seventy-seven years after it was published – that the Catholic Church placed *On the Revolutions* on its Index of Forbidden Books. From that point on books advocating heliocentricity were automatically relegated to the Index, a practice that only ended in 1758.

What had changed? Why, by the 1600s, had heliocentricity become a matter of life and death? What made it so dangerous that even the Church of Rome was running scared?

The answer to these questions lies almost entirely in the threat posed by one man . . .

CHAPTER TWO

THE HERMETIC MESSIAH

Although largely forgotten today, the Dominican monk-turned-heretic Giordano Bruno was regarded as one of the greatest intellects and philosophers of his time. The champion *par excellence* of the Hermetic tradition, he travelled Europe preaching its virtues and arguing for a root-and-branch reform of society based on its principles. He aimed to be Hermeticism's greatest prophet – even its messiah – but instead became its greatest martyr, ending his days in the searing embrace of the Inquisition.

Bruno was messianic, bombastic and stubborn, with a huge ego and belief in his own brilliance and importance. But then a man whose whole philosophy and mission in life centres on the Hermetic adage of *magnum miraculum est homo* is hardly destined to be a shrinking violet. He saw himself as living proof of just how miraculous a man could get. Where he parted company with most typical egocentrics, however, was that he considered all men and, less usual for the time, all women, as being either actually or potentially as brilliant as himself. The targets of his greatest fury were those who held people back, who told them they were insignificant and worthless. Surely it is difficult for a philosophy to be more diametrically opposed to the Christian doctrine of original sin, the idea that babies are born in a fallen state due to the famous transgression of Adam and Eve.

Bruno was first, foremost and totally besotted with Hermeticism, the great golden thread that connected his philosophy, religion and magic. He wrote a huge number of treatises and poems that contained coded and symbolic teachings, being heavily influenced by the works of Ficino and Agrippa, although characteristically he was never afraid to depart from them.

Bruno was born in 1548 – five years after the publication of *On the Revolutions of the Celestial Spheres* – in the town of Nola in the Kingdom of Naples, which comprised the whole of the southern half of Italy and, due to the complex geopolitics of the day, actually belonged to the Spanish king of Aragon. As we will see later, this area witnessed particularly odd activities during the sixteenth and seventeenth centuries, mostly concerning Dominican monks. Although he was baptised Filippo, when he became a monk in the Dominican monastery in Naples at the age of sixteen, he took the name Giordano (or 'Jordan', from the baptismal river). Like many bright kids from a humble background – his father was a soldier – his decision to become a monk was probably the only career move that allowed him to get an education. And he was indeed very bright, being particularly distinguished for his mastery of mnemonics and memory systems, even being summoned to Rome by Pope Pius V to explain how they worked.

The 'Nolan', as Bruno was often known, refused to let anybody tell him what to think or even what he could and couldn't study, which was something of a shortcoming in a sixteenth-century monk. In 1576, at the age of twenty-eight, he came under suspicion for heresy, or rather suspicion of suspicion of heresy. 'Suspect of heresy' was the formal term for a transgression against Church law, committed by those who read heresy and listened to heretics, even if they disagreed with them. At that time it was in fact best for one's health and safety to have no dealings with the work of heretics at all. (The official transgression had the somewhat Monty Pythonesque subdivisions of 'Vehemently Suspect' and 'Slightly Suspect', although there was nothing funny about the Vehemently Painful punishment.)

Though the details are a bit sketchy, it appears that all Bruno did was read and discuss ideas that had been condemned as heretical. He certainly debated the Arian heresy[1] in tones that weren't unequivocally negative and questioned the doctrine of the Trinity, largely because he thought it made no sense. (He later maintained to the Inquisition he had never denied the doctrine, only doubted it.) And he hid a copy of a book by the Dutch proto-Protestant Erasmus in the monastery toilet – although he could easily have explained away its presence as toilet paper, which would no doubt have appealed to his superiors. Perhaps that's what he did do. It would have been in keeping with his character.

Despite being mild compared to what he would preach later in his life, this string of actions coupled with his general freethinking was enough to attract suspicion, and so he abandoned the monastery and fled from Naples. For five years he wandered around northern Italy, southern France and Switzerland and appeared in Venice, Padua, Milan, Geneva, Lyons and Toulouse, among other places. Given the extent of his travels, it is impossible to pinpoint how and when Bruno became devoted to Hermeticism and magic. He may have started to study it in the monastery (perhaps in the toilet?), or perhaps encountered it during his wanderings, but the catalyst for his entrance into the world of the arcane is most likely to have been his fascination with memory systems.

The art of memory, which Bruno did much to help revive, developed in classical Greece as a system for storing and recalling information using specific mental images. So powerful is the system that it is still widely used today, even by celebrities such as the gifted British illusionist Derren Brown. However, an esoteric version of this technique that combined the mental images with magical principles could, it was believed, be used not just to remember what had already been learned but to acquire completely new information. Briefly, this version employed the principles of talismanic magic, in which different symbols, shapes, colours and materials are deemed to have specific properties and energies based on magical associations. The trick was to use those principles when forming the mental images. It was as if a portal opened and hidden knowledge flowed in. It was writing books on the magical art of memory that made Bruno's reputation when he settled in Paris in 1581, but by this time he had also developed some extraordinary ideas about the importance of magic in general and Hermeticism in particular – ideas which challenged its previously conceived limitations.

As we have seen, since the rediscovery of the Hermetica a century before Bruno's birth, many had believed Hermeticism was compatible with Christianity, as its sacred books could be seen to foreshadow the coming of Christ. However, as far as Bruno was concerned this line of reasoning didn't go far enough. As Frances Yates explains:

> Giordano Bruno was to take the bolder course of maintaining that the magical Egyptian religion of the world was not only the most ancient but also the only true religion, which both Judaism and Christianity had obscured and corrupted.[2]

Bruno burned with a sense of destiny, believing passionately that it was his mission to restore the old Egyptian religion, and that this would bring an end to Europe's political and social ills. He also saw Hermeticism as a way of transcending the religious schisms that were causing such horrors.

One of the keys to understanding his Egyptian passion is found in the famous section of *Asclepius* known as the 'Lament', in which Hermes warns of a time when the gods will abandon Egypt to the rule of foreigners, who will then establish their own false religions and ban the country's traditional faith on pain of death. This will, Hermes continues, be a tragedy not just for Egypt but for the world, since Egypt is the home of the gods on Earth, and once they leave the land they will be lost to all mankind. But, he goes on, in time the one God will intervene and the lesser gods will be restored, and 'stationed in a city founded at Egypt's farthest border toward the setting sun, where the whole race of mortals will hasten by land and sea.'[3]

Because the Hermetic books were believed to hail from the zenith of the Egyptian civilization, the Lament was seen as an authentically ancient prophecy. And since the time of their writing, it had come to pass that Egypt's native religion had indeed been eclipsed: since Alexander the Great's invasion in the fourth century BCE, the country had been under foreign domination – first Greeks, then Romans, then Christians and now Arabs. It stood to reason that if the first part of the prophecy was true, the second part might be also. The ancient gods might return, and a golden Hermetic city might be built that would draw the whole world to its magic.

While most considered the Greeks and Romans as the interlopers responsible for crushing Egypt's religion, Bruno singled out the Christians as the real villains. He may even have been right. Although the Greek and Roman overlords did import their own gods and cults, they also permitted the continued practice of religions native to the area. As we mentioned earlier, Clement of Alexandria witnessed processions of the Egyptian priests and priestesses, bearing the forty-two sacred books of Hermes, around the year 200. It was only when Christianity came to dominate in the fourth century that native Egyptian cults were ruthlessly persecuted and ultimately banned on pain of death. Bruno's interpretation of the Lament required no uncanny knowledge on his part, since Christian writings of the time recorded the suppression of Egypt's 'demonic' pagan cults with characteristic glee.

What excited and motivated Bruno most, however, was his conviction that the second part of Hermes' prediction – the restoration of the Egyptian religion and the return of the gods – would take place during his own lifetime. He interpreted the religious wars that were ripping Europe apart as the death throes of the faith that had suppressed the Hermetic religion. He also believed that Christianity was an offshoot of something much bigger and more ancient, despite it mistaking itself as the main event. Bruno did, however, admire the way of life Jesus taught, particularly the simplicity of the injunction to treat others as you yourself wish to be treated. (He seems to have regarded Jesus's mission as an attempt to take the Jewish religion back to its Egyptian roots, which our own research indicates to be at least partly correct.)[4] In a statement made to the Inquisition at the time of his arrest, Bruno is reported to have said, 'the Catholic religion pleases him more than any other, but that this too has need of great reform'.[5] He particularly deplored the way the Catholic Church sought to impose itself through 'punishment and pain'; using force rather than love to keep its worshippers was a sure sign that something was terribly wrong.

Yet even at its best, Bruno viewed Christianity as only an also-ran in the great race towards enlightenment and salvation. The ancient Hermetic religion of Egypt would soon assert its superior position when it returned to the

Earth through the mediation of its greatest prophet, Bruno himself.

Bruno believed that the great religious revolution on Earth would be preceded by upheavals in the heavens, reflecting the Hermetic principle (from the *Emerald Tablet*) of 'as below, so above/as above, so below'. Bruno moreover suggested an intriguing variation on this theme, namely that any changes would be echoed in a shift in mankind's *perception* of the heavens. And this, he believed, unlocked the true significance of the heliocentric theory.

For centuries the most learned of men had simply got cosmology wrong; Copernicus had shown that. But the Hermetic books, which Bruno believed preserved the most ancient wisdom of all, also stated that the sun was at the centre of all that mattered and that the Earth moved around it. Copernicus – who also invoked Hermes Trismegistus – had restored the correct perception of the order of the cosmos. Bruno thought Copernicus had proved mathe-matically what Hermeticists already knew but had never been able to prove. At the very least, he reasoned, establishing that the Hermetic philosophy contained demonstrable truths about the cosmos would surely win it more converts.

But Bruno also believed that Copernicus' work went way beyond vindicating the Hermetic treatises; he considered it as the key to the prophesied new Hermetic age. The fact that Copernicus had presented his proofs when he did was a portent of the coming changes. But not everybody had yet accepted the new system; it was still being hotly debated. If it could be established beyond doubt and enter into the canon of accepted fact, Bruno thought, then this would literally trigger the new age of Hermetic enlightenment. In turn, this would reveal a new way of comprehending the mysteries of creation, that is by using the intellect to obtain otherwise elusive proof of certain Hermetic magical and philosophical concepts, as summarized by Frances Yates:

> The marvellous magical religion of the Egyptians will return, their moral laws will replace the chaos of the present age, the prophecy of the Lament will be fulfilled, and the sign in heaven proclaiming the return of the Egyptian light to dispel the present darkness was . . . the Copernican sun.[6]

Ironically events showed that Bruno was at least half right. Establishing heliocentricity did indeed lead to a revolution that would change academic attitudes to religion, but it was the scientific revolution. The crucial Hermetic philosophy was simply lost along the way. Another cause to Lament.

THE MISSION

It was no accident that Bruno decided to begin his mission in Paris. The city was the perfect place given that the centre of the Renaissance had shifted to France as the sixteenth century unfolded (neatly symbolized by Leonardo da Vinci's own move to France at the invitation of the king in 1510).

This shift was a consequence of the Catholic Church's attempts to reverse the damage of the Protestant Reformation, through their Counter-Reformation. This was kicked off by the Council of Trent, initiated by the Pope in 1545 – and which continued for eighteen years – to tighten up and rigidly define Catholic doctrine and practices. One result of the Council was that the Church came to assert greater control over the arts, which included, for example, the banning of non-Christian, and especially pagan, imagery in paintings and sculpture. (No more depictions of Isis and Hermes by popes. Amazingly those in the Appartamento Borgia were allowed to remain.) These prohibitions bit more deeply in Italy than in France, where the Church's real power over French daily life might best be summed up by the timeless Gallic shrug. As the cultural centre of the Renaissance had relocated to Paris, it also became a great centre of Hermeticism, even among Catholic scholars and intellectuals. Both developments owed much to the sophisticated thirst for knowledge of the French court.

Although of course outwardly Catholic, King Henri III of France was a devotee of the occult philosophy. The celebrated poet and chronicler Agrippa d'Aubigné recorded how after swearing him to silence, Henri had revealed a collection of magical treatises he had had brought in from Spain. In this he was only maintaining the family tradition, since his mother was Catherine de' Medici, the great-granddaughter of Lorenzo the Magnificent. Then in her sixties, she still exerted a powerful influence of her own in Paris. Very much a de' Medici, like her ancestor the great Cosimo, Catherine was a renowned patron not just of the arts but also of astrologers and magicians. So it was hardly surprising that Henri III, the third of her sons to reign in France, shared her arcane interests.

But Henri was, from Bruno's perspective at least, also ideally positioned in Europe's power politics, in which a major conflict between the Catholic and Protestant nations was looming. Henri had a relaxed attitude to Protestantism, both at home and abroad, and as a result of anxiety about the strength of the major Catholic power of Spain he favoured closer ties with Protestant England, Spain's great enemy. Many, not just Bruno, saw Henri as Europe's best hope for a peaceful and tolerant future. As a powerful Catholic monarch with a zealous interest in magic and the Hermetica and no animosity towards Protestants, Bruno considered Henri the ideal leader of his Hermetic revolution. There are indications in other books published in Paris at that time, and in plays being performed in the king's honour such as the *Ballet comique de la Reine* (the first ballet, staged for the court of Catherine de' Medici in 1581) that Bruno was not alone in this view of Henri.

Meanwhile a well-established circle of expatriate Italians who had settled in Paris because of their heterodox ideas (probably because of the Medici influence) welcomed Bruno with open arms. More significantly, these Italians had some influence over the king. But lurking behind the Franco-Italian circle was, inevitably, an *eminence grise*, a secret adviser and friend of the greatest movers and shakers of the time. This shadowy force-to-be-reckoned with was one Gian Vincenzo Pinelli of Padua (1535–1601), a scholar and collector (primarily a botanist but his interests were truly Renaissance in scope and depth) best remembered today as Galileo's mentor. Pinelli had built up a pan-European network of correspondents and informants who reported to him on not just scientific and cultural issues but also political events. Unsurprisingly, he therefore showed great interest in Bruno's arrival in Paris and they are likely to have met when Bruno visited Padua during his wanderings.

After the larger-than-life Hermeticist arrived in the French capital in 1581, he gave public lectures and published two books on the magical art of memory. Bruno soon attracted the attention of the King, and having cannily dedicated the first of his books, *On the Shadows of Ideas* (*De umbris idearum*) to Henri, was duly summoned for a royal audience. As a reward he was given a paid lectureship at one of Paris' colleges. His next move was more surprising: in the spring of 1583 he left Paris for London, where he was to spend more than two years and produce his most important work. The English ambassador in Paris sent a report to Queen Elizabeth's spymaster, Francis Walsingham, advising him of the impending arrival of Bruno, 'whose religion I cannot commend'.[7] With a nice ironic edge Bruno described himself to the Oxford scholars as a 'doctor of a more abstruse theology'.[8] Well, yes. That's one description of it.

Although he had no official diplomatic standing, Bruno was clearly on some kind of unofficial, or semi-official, mission to England. Travelling with letters of introduction from Henri, he lived in the house of the French ambassador, Michel de Castelnau, Sieur de Mauvissière. Because he kept such close company with Castelnau – even accompanying him regularly to Queen Elizabeth's court – and Castelnau was in turn happy to be known as an associate of Bruno, it fostered the impression that the latter had the French king's backing. And it seems Henri had no problem with that.

As to the purpose of Bruno's mission, it fitted perfectly his agenda of uniting Christianity and averting a catastrophic war in Europe. The idea was to get the Catholic nations to band together under a single monarch and the Protestant nations to unite under another, both of whom would be advised and influenced by Hermeticists who would ensure peace between them. Henri III and Elizabeth I were prime candidates.

English esoteric circles, too, had great influence at the royal court, most obviously in the shape of John Dee (1527–1608), Elizabeth's astrologer and adviser in many areas, including diplomacy, espionage and the expansion of English influence across the globe. Although there is no record of Bruno and Dee meeting, because they had mutual friends and frequented the same court and intellectual circles they almost certainly did. This was especially likely as Dee was not only a champion of the Copernican theory but also a passionate devotee of the Hermetic tradition.

Bruno met the Queen herself on the many occasions he accompanied Castelnau to court, declaring himself a fervent admirer of the 'diva Elizabetta' and proclaiming her superior to any man in her heroism, learning and wisdom.[9]

That fulsome compliment 'diva' was to count against him with the Inquisition, since they took against calling a declared heretic 'divine'. Worse by far, Elizabeth was an illegitimate heretic in Catholic eyes at least. In any case she was female. And she had a sure sign of the witch, being red-haired. But Bruno enthusiastically joined the cult of the Virgin Queen, which lauded her as the potential spearhead of a new age, the bejewelled goddess who would unify Protestant Europe. He seems to have admired the relatively peaceful nature of Elizabethan England when compared to the internal divisions that were then tearing apart the other nations of Europe.

The uncompromising Neapolitan took part in a famous debate with the scholars at Oxford, in front of the Polish prince Albert Laski and the eminent courtier and poet Sir Philip Sidney, in which he endorsed Copernicus' ideas and linked them to magical concepts about the sun derived from Marsilio Ficino's work.

It was in England that Bruno wrote some of his most important books. Of these, all apart from the first were penned in Italian rather than the customary Latin. But why go to London to publish books in Italian? Of the few Londoners who could read in the first place, how many could read Italian? Presumably Bruno's books targeted Italians in London and Paris, a readership who would then take his ideas back to their homeland. Or perhaps Bruno had intended that the books be shipped over to Italy? Either way, they were circulating there within a few years, as we will see.

The first – and only Latin – work he published in his first year in London was *Explanation of the Thirty Seals* (*Explicatio triginta sigillorum*), a book about the magical memory system that culminates in an essay about the Hermetic vision. In this, Bruno lists Moses and Jesus as among those who had achieved enlightenment through this means. The latter is portrayed not as the Son of God, or even as a divinely appointed prophet, but as a gifted and advanced magus, a practitioner of the same art so beloved by Bruno. This is an interesting concept – the founder of the religion that saw Bruno's work as heretical practising the same heresies himself – but one that is not without some foundation, as we have discussed elsewhere.[10]

In 1584 Bruno published two key works, both of which relate to Copernicus and heliocentricity. The first was *The Ash Wednesday Supper* (*Cena de le ceneri*), a dialogue between a group of scholars as they journey around London. In this book Bruno praises Copernicus, although he also claims that even Copernicus never came to understand the full importance of his discoveries. With his usual bravura, Bruno also declares himself to be Copernicus' heir and states his intention to use his revelations to free the human spirit.

The second book was *Expulsion of the Triumphant Beast* (*Spaccio della bestia trionfante*) a 'glorification of the magical religion of the Egyptians',[11] an unequivocal declaration of the need for its return in order to restore balance to the world. He links this to the Lament in *Asclepius*, which he reproduces in full.

The drama of the *Triumphant Beast* takes the form of a gathering of Greek and Egyptian deities to reform the heavens, changing the constellations in order to produce a similar shift on Earth. This is modelled on the Hermetic treatise *The Virgin of the World* (*Korè Kosmou*) in which Isis describes a similar council of the gods to her son Horus. She also features, alongside Sophia, in Bruno's work. The 'triumphant beast' is, according to Bruno's dedication to Sir Philip Sidney, the sum of all the vices that prevent human beings from activating their divine potential. However, some – including, fatally, the Inquisition – interpreted it as a veiled reference to the Pope. A political subtext runs through the *Triumphant Beast*, as it ends with the council of the gods praising the great virtues, pureness of heart and magnanimity of Henri III, and his fitness to preside over a spiritually unified Europe.

Another significant work Bruno wrote and published in London in 1585, also dedicated to Sidney, was *On the Heroic Frenzies* (*De gli eroici furori*). Ostensibly a collection of love poems, it soon becomes clear that the 'frenzy' of passionate love is a way of attaining the Hermetic gnosis. This concept is taken from Agrippa (in turn a development from Ficino), who wrote of four types of *furor* that enable the soul to reconnect with the divine: poetic inspiration, religion, prophecy and love, the *furor* of Venus. Of the last, Agrippa writes that it 'transmutes the spirit of a man into a god by the ardour of love, and renders him entirely like God, as the true image of God'[12] before proceeding to cite Hermes Trismegistus, from *Asceplius*, as an authority for this idea. This is obviously why the idea was so attractive to Bruno.

The concept of erotic love as a portal to Hermetic illumination links Bruno with other well-established traditions of sacred sexuality, including sex magic and tantrism. For someone who elevated what we would now call the sacred feminine, and who admired intelligent and able women, it is curious that nothing in the historical records specifically links him with any women. Or man for that matter: if Bruno had even been remotely rumoured to be gay this would have featured in the Inquisitions list of his calumnies. As it is, the Inquisition records only suggest that he was a womanizer, without any actual proof.

Bruno wrote in his dedication to Sir Philip Sidney that, although he hadn't had as many lovers as Solomon, it wasn't for the lack of effort on his part:

I have never had a desire to become a eunuch. On the contrary I should be ashamed if I agree to yield on that score were it only a hair to any man worth his salt in order to serve nature and God.[13]

Only one source links Bruno, if only obliquely, with affairs of the heart. Several historians have suggested that the character of Berowne, the leader of the poets at the court of the King of Navarre in Shakespeare's romantic comedy *Love's Labour's Lost* is based on Bruno. The identification is highlighted, as Yates has shown, by the fact that some of Berowne's speeches, particularly his great paean in praise of Love in Act IV ('For valour, is not Love a Hercules . . . '), contain specific parallels to *Expulsion of the Triumphant Beast*, the greatest of the works Bruno wrote in England, about ten years before Shakespeare penned the play.

Love's Labours Lost is not one of Shakespeare's most popular works because of its abstruse and often tedious wordplay. The plot describes the oath taken by the King of Navarre and three of his scholars, led by Berowne, in order to concentrate on their pursuit of knowledge, which entails living an abstemious life for three years, including forswearing the company of women. But the arrival of the Princess of France and a bevy of young ladies-in-waiting throws several cats among the pigeons, with predictably hilarious(-ish) consequences. Other than the lesson that locking oneself away in the pursuit of knowledge is a bad idea – wisdom comes from participating in the real world – there seems little message in this typically mannered Elizabethan romantic comedy. Most of the jokes have never been found funny since doublet and hose went out of fashion.

But there is a bit of a mystery surrounding *Love's Labours Lost*. The play has no proper ending – all of the characters simply disperse with a promise to meet up again in a year's time. There are also a couple of contemporary references to an otherwise unknown sequel by Shakespeare called *Love's*

Labour's Won, but for some reason this has been omitted from the Shakespeare canon that passed into history. One clue, however, lies in the fact that at the time the play was written the King of Navarre and the King of France were one and the same, and he was being supported by Bruno and other Hermeticists – as we will see.

(However, at least one good thing came out of this little literary mystery. It inspired the 2007 *Dr Who* story 'The Shakespeare Code', in which David Tennant's Time Lord discovered that the now-lost *Love's Labour's Won* contained coded magical utterances that were set to open a portal to another dimension.)

THE INFINITE UNIVERSE

In addition to his zeal for Hermetic reformation, Bruno was unquestionably one of the greatest intellects of his time, and was especially admired for his scientific and mathematical ideas and theories. Several studies have been devoted to this side of him, including Paul-Henri Michel's *The Cosmology of Giordano Bruno* (1962), Dorothea Waley Singer's *Giordano Bruno: His Life and Thought* (1950) and Hungarian academic Ksenija Atanasijevic's *The Metaphysical and Geometrical Doctrine of Bruno* (1923). Atanasijevic describes him as 'certainly the greatest philosopher of the XVIth century',[14] and writes:

> If the Inquisition had not managed to put its jackal's claws upon him when he was forty-four and if he had not been burnt alive at the age of fifty-two, Bruno would have left to humanity some more of his inspired and farsighted conceptions.[15]

Many of his pronouncements – all derived from the essential principles in the Hermetica – were staggeringly ahead of their time.

Clearly in a fever of composition, while still in London in 1584, Bruno published another remarkable work: *On the Infinite Universe and Worlds* (*De l'infinito universo e mondi*), in which he proposed two ideas that went way beyond even those of Copernicus. The first was that all creation was not contained within the space bounded by the sphere of the fixed stars, but was infinite. The second was that the stars are not small bodies of light fixed on that sphere but are actually suns like our own, only immensely far away, at different distances in the infinite universe. Bruno made a further extrapolation: if the stars are suns, then they too are circled by planets. He wrote:

> For there is a single general space, a single vast immensity which we may freely call *Void*; in it are innumerable and infinite globes like this on which we live and grow. This space we declare to be infinite, since neither reason, convenience, possibility, sense-perception or nature assign to it a limit. In it are an infinity of worlds of the same kind as our own . . . Beyond the imaginary convex circumference of the universe is Time.[16]

The last sentence is strangely prescient of the curvature of space-time that is regarded as one of Einstein's greatest insights.

Not only did Bruno think there were other planets, but also that some were inhabited. *On the Infinite Universe and Worlds* takes the form of a dialogue between two characters, Fracastoro and Burchio. At one point, the latter asks whether the other worlds are inhabited like ours, to which Fracastoro replies:

> If not exactly as our own, and if not more nobly, at least no less inhabited and no less nobly. For it is impossible

that a rational being fairly vigilant, can imagine that these innumerable worlds, manifest as like to our own or even more magnificent, should be destitute of similar or even superior inhabitants.[17]

Ideas such as the one expressed by Fracastoro are so extraordinarily modern that it is difficult to appreciate just how big a conceptual leap they were at the time – and just how shocking they could seem.

Even Copernicus had maintained the conventional idea of a fixed sphere of stars. As such, shifting the centre from the Earth to the sun made relatively little difference to established views of mankind's special place in creation. Even though the Earth was no longer the centre of everything, the sun is, making mankind *almost* the focus of creation. And according to Copernicus there was still only one relatively small, finite cosmos, in which existed a singular world in which God had created living things: a cosmos made just for us.

But if there are other suns, with their own inhabited planets, then the unique specialness of this world and of humanity is called into question. Since an infinite universe can have no centre, neither the world, nor even the sun, could claim to fill this role. In this theory of the world, humankind is shifted further from the centre of things – and from being the focus of God's creation.

Modern science, which emphasizes the insignificance of both humanity and the Earth in cosmic terms, credits Copernicus with beginning the shift in perception from humanity being the centre of everything to our inhabiting a tiny part of an infinite universe. However, the credit should really belong to Bruno, since it was his notion of an infinite universe that provided the truly radical leap.

There was one major and insurmountable difference between the modern view and Bruno's. He would never have accepted the twenty-first century reasoning that, because the universe is infinite and we are not alone in it, human beings are therefore unimportant. He believed that the universe teems with life, including us, because it was *made* for life.

Another major difference between Copernicus' and Bruno's cosmologies was that Bruno's unequivocally clashed head-on with Christian teaching, flatly contradicting the biblical story that God created the sun, moon and stars after making the Earth, with no mention of other earths. One of the heretical ideas for which Bruno was executed was that of an infinite, inhabited universe. So what was the source of his radical ideas?

In fact, Bruno derived the notion of an infinite universe from a passage in *Asclepius*, in which Hermes refers to a region 'beyond heaven', which implies that the heavens are not bound by the sphere of the fixed stars.[18] Although this suggests an infinite universe, it does not state that it is full of suns. The idea therefore seems to have been Bruno's own extrapolation.

As we noted in the last chapter, at least one thinker had challenged the 'celestial sphere' concept and argued for infinite space. This individual was Englishman Thomas Digges, 'the first Copernican in England',[19] whose ideas Shakespeare alluded to in the 'nutshell' line in *Hamlet*. Digges made the proposal in 1576 in his outline of Copernican theory – the first published in England – *A Perfit Description of the Caelestiall Orbes*. Given that Bruno wrote his work in England, it could be that he was influenced or inspired by Digges.

But Digges, too, was part of the Elizabethan esoteric scene, being a protégé of John Dee, himself a great supporter of heliocentricity. Although Dee left no reference to the theory in his own works, he encouraged its first champions in England, urging the astronomer John Field to use Copernicus' system to draw up a table of the positions of the planets in 1557. Dee was also, notably, Digges' mathematics tutor (Digges called Dee his 'second mathematical father').[20] In fact, Digges' version comes straight from *Asclepius*.[21]

These were not the only anticipations of modern scientific thinking and discoveries in Bruno's work. In fact, some of his ahead-of-their-time pronouncements become positively eerie. In *On the Infinity of the Universe and Worlds* he writes:

> Thus soul and intelligence persist while the body is ever changing and renewed part by part . . . for we suffer a perpetual transmutation, whereby we receive a perpetual flow of fresh atoms and those that we have received are ever leaving us.[22]

As we now know, every cell in our bodies is constantly being replaced throughout successive cycles of seven to ten years. But how did Bruno know? And that is by no means the limit of Bruno's prescience. Peter Tompkins writes:

> The doctrine of evolution, the progressive development of nature, an idea unknown to classical philosophy, was first pronounced by Bruno, not vaguely or partially; he extended its laws to the inorganic as well as the organic world, maintaining that unbroken line of evolution from matter to man which only modern science later began to recognize.[23]

Bruno heavily influenced the English natural philosopher and physician William Gilbert (1544–1603) who British science writer John Gribbin describes as 'the first person to set out clearly in print the essence of the scientific method – the testing of hypotheses by rigorous experiments – and to put that method into action.'[24]

Gilbert's major work, *On the Magnet, Magnetic Bodies, and the Great Magnet of the Earth* (*De magnete, magneticisque corporibus, et de magno magnete tellure*), published in 1600, was one of the landmarks of the scientific revolution, presenting his theory that the reason magnets, or loadstones, work is because the Earth itself is a magnet. Historian Hilary Gatti, author of a study of Bruno's legacy to England following his visit, demonstrates that in his ideas about the Earth's magnetism, Gilbert built on Bruno's cosmology.[25]

A collection of Gilbert's papers published half a century after his death, *A New Philosophy of Our Sublunar World* (*De mundo nostro sublunari philosophia nova*), makes his debt to Bruno very clear.[26] The two men almost certainly met, as Gilbert was physician to Elizabeth I at the time that the Neopolitan was a frequent visitor to her court.

Another royal physician who made an indelible mark on the history of science was William Harvey, who as Charles I's physician in 1628 famously demonstrated the circulation of the blood – 'one of the greatest achievements of the Scientific Revolution'.[27] However, as Harvey acknowledged, his inspiration came from the work of one of his colleagues, the Hermeticist Robert Fludd (who we will meet in a later chapter), who had proposed the idea based on Hermetic principles. Fludd's own inspiration was almost certainly his esoteric hero Bruno, who had put forward the same thing for the same reasons nearly half a century earlier.[28] Once again, he deduced this from the Hermetica, specifically its association of the spirit that moves through the body with the blood; Treatise X of the *Corpus Hermeticum* explicitly states 'the spirit, passing through veins and arteries and blood, moves the living thing'.[29] And so another major scientific discovery can be atributed to Hermes Trismegistus – and to Bruno.

His influence was, indeed, vast. As Ksenija Atanasijevic writes:

But Bruno's contribution to the development of subsequent philosophy and modern astronomy is beyond proper evaluation not only in terms of his conception of the infinity of the universe; with his comprehensively conceived and elaborately argued doctrine of the triple minimum he is also one of the leading forerunners of later monadology, atomism and the teachings about the discontinuity of space, time, motion and geometrical bodies.[30]

Atanasijevic concludes that 'it was Bruno who laid the firm foundations upon which was to rise, in the course of time, the . . . edifice of new atomic science'.[31] But although Bruno's ideas were in many respects far closer to the modern scientific mindset than the works of Copernicus and Galileo, they sprung from his immersion in the ancient philosophy of Hermeticism.

THE GIORDANISTI

Bruno returned to Paris with Castelnau in the autumn of 1585, being attacked by pirates as they crossed the Channel – much like Rosencrantz and Guildenstern in *Hamlet*. Things were fraught in Paris: a group of ultra-Catholic French nobles had formed the Catholic League, which aimed to oust Henri III and wipe out the French Protestants – the Huguenots – and form an alliance between France and Spain. Henri had been forced to make a number of concessions such as rescinding liberties he had granted to the Huguenots, in order to avoid civil war. Henri had no heir and France was simmering with tension as sides were being taken over who would succeed him.

Somewhat surprisingly, in Paris Bruno made overtures to the papal nuncio about returning to the Catholic Church and receiving absolution, although he was spurned. This seems incongruous, but Yates explains that Bruno had become convinced that the great Hermetic reformation would happen *within* the Catholic Church, so that was the place to be. As she wrote, 'The new dispensation was to be an Egyptianized and tolerant Catholic and universal religion, reformed in its magic and reformed in its ethics.'[32]

However, it soon became apparent that this rather unrealistic hope was doomed, with political events in France taking a turn for the worse for Bruno's programme of reform. He left Paris in the late summer of 1586, shortly before the Catholic League took control of the city. Adapting himself to the new situation, Bruno shifted his focus to the Protestant lands, and toured Germany for the next few years. Initially he obtained a post as lecturer at the University of Wittenberg in Saxony (which had produced Martin Luther, not to mention the fictitious Hamlet). Bruno owed his job to the influence of another important Oxford contact, Professor of Law Alberico Gentili, an Italian refugee whose family had fled abroad because of their Protestant beliefs. Gentili is remembered today as the founder of international law.

After a couple of years at Wittenberg, Bruno moved on briefly to the Prague court of the Holy Roman Emperor Rudolph II. Despite his leading role in the great Catholic dynasty of the Habsburgs, Rudolph (1552–1612) was extraordinarily liberal-minded. Not only was he renowned for his patronage of the arts and learning but he was also an active and enthusiastic sponsor of the occult sciences, particularly alchemy. Rudolph employed Tycho Brahe as his Imperial Mathematician, who was himself succeeded by his assistant, Johannes Kepler. Shortly before Bruno arrived at his court, the great Dr Dee had been a distinguished guest of the Emperor.

Rudolph never shared his dynasty's political or religious interests, and focused instead on his own enlightened pursuits. He moved the imperial court from Vienna to Prague in Bohemia, which under his patronage became a sparkling Renaissance city, where all learning and culture was encouraged. In Prague, Protestants and – extraordinarily for the time – Jews were free to practise their religion. Rudolph also worked for a unified Christian Europe, backing those who worked for tolerance and reconciliation between Catholic and Protestant. His own religious orientation is unclear. Although raised a Catholic, he was obviously lapsed, going so far as to refuse the last rites on his deathbed. But neither did he join any of the Protestant churches.

Rudolph acted like a magnet for occultists, artists and scholars, and Bruno was no exception. But to Bruno an added attraction must have been the existence of a court of exceptional tolerance and open-mindedness. Having received some financial assistance from the Emperor, Bruno moved swiftly on to the University of Brunswick, all the while in a ferment of thinking and plotting.

Throughout his wandering years, Bruno's position on the Catholic Church and the nature of the Hermetic revolution shifted. Until his departure from Paris, he believed that an Egyptian reformation could begin within the Church, through collaboration between Hermes-friendly monarchs such as Henri III and allies in Rome itself. But not only was Henri losing the civil war against the Catholic League, he was soon to be assassinated by one of their agents, a Dominican monk. (Catherine de' Medici also died – surprisingly of apparently natural causes – at the beginning of that year.) Spain was bringing its whole might to bear on crushing Bruno's next best hope for harmony in Europe, Elizabeth's England, building up the armada for the attack of 1588; few gave England much of a chance.

At this time, when Catholicism seemed on the brink of triumph, a strangely symbolic event took place in Rome. In 1586 a great ancient Egyptian obelisk that had remained neglected for over a thousand years was moved to the centre of St Peter's Square. During the Roman Empire, many obelisks and statues were carried off to the imperial hub from Egypt and erected around the city, usually in honour of some emperor or another. Unsurprisingly, they had been knocked over and vandalized as nasty pagan monuments when Christianity became the state religion, but many were left where they fell, either in pieces or whole, to disappear beneath the ground over the centuries. In the sixteenth century only one obelisk was still standing, albeit with its base deeply buried, in a dingy alley behind St Peter's. Nearly three thousand years old, it had been taken to Rome on the orders of Caligula.

In 1586 Pope Sixtus V ordered that the obelisk be moved to its prominent place and following a monumental engineering effort that stretched the resources and skills of the day to their very limit, this 83-foot-tall (25-metre), 350-ton monument stood tall in the centre of the square. After being duly exorcised, it was topped with a large iron cross and had inscriptions honouring Christ (and of course Sixtus) carved into it.

Sixtus' declared motive was to assert the triumph of Catholic Christianity over paganism and to 'eradicate the memory of the superstitions of antiquity by raising the greatest footing ever for the Holy Cross'.[33] At first glance, this seems rather strange, since Christianity had put an end to paganism long before and the major threat to Catholicism at the time was Protestantism. But in the context of the Hermetic, Egyptian undercurrent the desire of this ultra-conservative and reactionary ex-Inquisitor – of whom it was said that he wouldn't even forgive Christ of his sins – to symbolize his Church's superiority over Egypt certainly makes sense.

For his part, Bruno became much more confrontational, publicly denouncing the Catholic Church and the Pope as both tyrannical and the cause of disorder and violence in Europe. He also changed strategy and decided that the Hermetic revolution would now be brought about by stealth, using more clandestine methods. He devoted much of his time in Germany to organizing a secret society, the Giordanisti, to further his ambitions. This underground network would act as contingency should there be a Catholic take-over of Europe, which seemed only too likely. The Giordanisti were effectively a Hermetic resistance movement. One fellow guest of the Inquisition in Rome said that Bruno had declared:

> . . . that he had begun a new sect in Germany, and if he could get out of prison he would return there to organize it better, and that he wished that they would call themselves Giordanisti.[34]

The chief informer against him, Zuan Mocenigo, said that shortly before his arrest Bruno had 'revealed a plan of founding a new sect' to him.[35] Although this revelation suggests that Bruno was still at the initial planning stages, his activities just before returning to Italy suggest otherwise. In retrospect it seems improbable that such a messianic figurehead would *not* have organized cells of disciples wherever he went, linking them into an underground network. Forming secret groups is what Hermeticists do.

Bruno had certainly acquired disciples and devotees in France and England. During his return to Paris he published works under his followers' names in order to disguise his authorship – although this may not have been favourable for those whose names he adopted – another sign that he was becoming more cautious and secretive. He was now building a following in the states of Germany. And despite restrictions caused by the problems of transport, because the formal organization was university-based, there would have been a constant movement of professors and students to other parts of Europe, all carrying Bruno's message.

Part of Bruno's new project involved the publication, in 1590 and 1591, of three lengthy poems expounding his magical philosophy, the progress of which he controlled more meticulously than any of his more overtly arcane and philosophical works. He even travelled to Frankfurt to oversee their production. One of the poems, *On the Threefold Minimum and Measure* (*De triplici minimo et mensura*) included symbols and diagrams for which – uniquely – Bruno made the woodcuts himself.

It has been suggested that Bruno lavished all this love on this particular work because it incorporated the Giordanisti's secret symbols and contained ciphered messages for its initiates.[36] Again, this makes sense in terms of a feared Catholic clampdown, in which his overtly Hermetic treatises would be banned. Of all Bruno's works this was the one that was ultimately responsible for his downfall.

Being such a high-profile possessor of Hermetic secrets was never going to be a passport to freedom of speech and a guarantee of personal safety, but clearly something in Bruno's character either persuaded him he would always lead a charmed life or he simply craved danger. Perhaps he also craved martyrdom.

A fiery fate was already waiting in the wings. While in Frankfurt, Bruno met Giovanni Battista Ciotto, an innocent-seeming book dealer from Venice. Back home, Ciotto sold a copy of Bruno's poem *On the Threefold Minimum and Measure* to a wealthy Hermetic dabbler, Zuan Mocenigo, which prompted him to invite Bruno to be his guest and teacher. At the age of forty-three, and after ten years away from Italian soil, Bruno accepted the offer. This would not turn out to be his best idea.

To modern eyes it seems as if Bruno was somewhat over-optimistic, seeing his return to Italy as a golden opportunity to inveigle himself into the Pope's favour. He even wrote to an old Dominican acquaintance in Venice that he hoped to receive papal absolution. Certainly further political change had rekindled his hopes of establishing a new age of Hermetic religion through an internal transformation of the Catholic Church. He still envisaged a French monarch who would bridge the divide between Catholics and Protestants, but fate would ultimately act against him there, too.

In the struggle over the succession that had followed the assassination of Henri III, another Henri, the King of Navarre, had triumphed (with the aid of English soldiers sent by Elizabeth). Navarre was a kingdom in southern France, on the Atlantic coast, the remnant of a larger and once predominantly Spanish kingdom that had straddled the Pyrenees. In 1589 the Huguenot king of Navarre also became King of France. In a politically expedient move, the new Henri IV converted to Catholicism, but as an ex-Huguenot it was widely anticipated he would unify the religious divide in France. Curiously and probably not coincidentally, he had his marriage annulled and married a Medici, Marie, daughter of Francesco de' Medici. Hermetic hopes once centred on Henri III now segued onto Henri IV. Bruno went so far as to tell his Inquisitors that he hoped that the new king would 'confirm the orders of the late King' (Henri III) for the favours granted to him.[37]

Bruno's sense of destiny had also been bolstered by other events, and without the grim knowledge provided by hindsight, perhaps it is easy to understand how he might have been so tragically misled. In 1591, Francesco Patrizi, Professor of Philosophy at the University of Ferrara, published a new edition of the Hermetica. In his dedication to Pope Gregory XIV, Patrizi urged him to decree that Hermetic philosophy be incorporated into the heart of

Catholic education. Gregory died soon afterwards, but his successor, Clement VIII, rewarded Patrizi for his efforts by bestowing him with the Chair in Platonic Philosophy at the University of Rome. Bruno told Mocenigo that he had taken heart from this, and expected the same kind of treatment from Clement. There was, however, a major difference. Patrizi was advocating the incorporation of Hermeticism into Catholicism, not vice versa like Bruno. And, of course, while ostensibly rewarding Patrizi – or perhaps buying him off – Clement never actually acted on his proposition.

It was in this climate that Bruno accepted Mocenigo's suggestion to travel to Venice. Accompanied by his secretary, Jerome Besler, Bruno initially declined Mocenigo's invitation of hospitality, and stayed in his own lodgings. He gave talks at Ciotto's bookshop and frequented intellectual salons in private homes, besides spending three months at Padua, hometown of the *eminence grise* Gian Vincenzo Pinelli, whom he undoubtedly met. Only in the spring of 1592 did he finally give in and agree to stay with Mocenigo. During his two-month visit his host made notes of their conversations, which no doubt seemed innocent enough, perhaps even flattering at the time, but they were to provide the basis of the case against him.

There were other good reasons why Bruno and his network wanted to shift their focus to Venice. The republic was becoming a centre of opposition to the Pope's authority and there were moves to forge a political and religious alliance with England (although this only gathered momentum in the years after Bruno's death). Astonishingly there were even hopes that Venice might adopt Anglicanism, which probably explains why the Pope excommunicated the whole republic in 1606. The key figures in this plan were all associated with Bruno. They included the English ambassador (and former spy) Sir Henry Wotton, who had been at the Italian's controversial lecture in Oxford and was a great friend of Alberic Gentilio, the professor of law who facilitated Bruno's career in Germany. Another was Traiano Boccalini, author of *News From Parnassus* (*Ragguagli di Parnaso*), which, modelled on *The Expulsion of the Triumphant Beast*, called for a 'general reformation of the whole wide world'.[38]

The unravelling of events such as these in Venice and Padua (part of the republic of Venice) in the aftermath of Bruno's visit was unlikely to have been coincidental. Neither was it much of a coincidence that Padua appears to have become a sudden magnet for Hermeticists when Bruno left.

And then, suddenly, it all became too obvious. In May 1592, when Bruno was preparing to return to Frankfurt, Mocenigo refused to let him leave, hiring a gang of gondoliers to lock him in a room, and sent for the Inquisition. Bruno was to be their prisoner for the remaining eight years of his life, with the resulting agonizing ending usually reserved for those who spoke out against ignorance and tyranny.

No evidence remains to suggest why Mocenigo decided to play the villain. Some believe his invitation was a trap from the start, or even that he had been in the pay of the Inquisition from the moment he bought *On the Threefold Minimum and Measure*. Others think that Mocenigo's enthusiasm for Bruno's philosophy was genuine but that he became disillusioned or alarmed. Perhaps Mocenigo simply feared for his immortal soul.

Bruno was questioned by the Inquisition and then tried in Venice. The major concern was the 'great reform' he preached. He did recant his heresies and begged for mercy from the judges, but the Supreme Inquisitor in Rome sent for him. Bruno was kept in prison in Rome for five years without so much as being questioned. After finally being interrogated, he was kept imprisoned for a further three years, without being tried. Heretics who admitted their errors – as Bruno appears to have done – were generally either given a prison sentence or released, albeit with restricted movements. Those who didn't were tried and, if found guilty, imprisoned or executed. Either way, a prisoner was generally dealt with relatively swiftly. Why the Inquisition dithered over Bruno is a puzzle, although we can offer a possible explanation that relates to the Hermetic undercurrent.

The inexorable endgame for Bruno finally began with the arrival of the newly appointed Cardinal Inquisitor Roberto Bellarmino (1542–1621, canonized in 1930). One of the most formidable intellects of the Church, Bellarmino was a loyal and capable pair of hands trusted by a succession of popes. He was a member of the Society of Jesus – another prong of the Counter-Reformation formed some sixty years earlier. The Society, known commonly as the Jesuits, was and is a notoriously unsentimental brotherhood, zealously committed to the unswerving maintenance of Catholic doctrine. Bellarmino's speciality was combating heresy, about which he knew a great deal, having taken infinite pains to comprehend the mindsets and arguments of heretics (although in his case, studying the subject was unlikely to see him accused of being suspect of a suspicion of heresy). A fierce and clever polemicist, he even engaged in a pamphlet war with James I of England.

Bellarmino had been an assistant to the papal emissary sent to negotiate with the Catholic League over the successor to Henri III after they assassinated him, negotiations that were trumped by the accession of Henri of Navarre. So he was aware of the Protestant and Hermetic expectations centred on the French kings.

When Pope Clement VIII appointed Bellarmino Cardinal Inquisitor in 1599, he reopened proceedings against Bruno, who asked that he be allowed to write a petition to Pope

Clement VIII declaring that he was prepared to defend the beliefs he was charged with, but that if Clement proclaimed them to be heretical, he would abide by his decision. Bellarmino didn't even show the petition to the Pope. According to the Cardinal Inquisitor, when Bruno was presented with a list of specific heresies in his work he abjured them, but then later withdrew this admission. This, as Bruno must have known, was the worst thing he could have done, as the most severe sentences were reserved for relapsed heretics. It was inevitable that he would be burnt at the stake. So had Bruno really changed his mind? No one will ever know. When he was led out to the pyre, his tongue was tied to prevent him speaking.

The record of the prosecution in Rome was lost after being taken to Paris in 1810 with the papal archives on the orders of Napoleon. However, we discovered from a summary of the Roman Inquisition's evidence found in 1942 (among the personal papers of the nineteenth-century Pope Pius IX) that Bruno was condemned for holding opinions contrary to the teachings of the Catholic Church, in particular about the Trinity, Jesus' divinity and tran-substantiation and speaking out against the Church; denying Mary's virginity; practising magic and divination; and claiming that there were many worlds in an infinite universe, and that the Earth moved. The German scholar Caspar Schoppe, who witnessed the execution, listed the heresies for which Bruno was being burned. These included the belief that there are innumerable other worlds; the promotion of the practice of magic; the claim that the Holy Spirit and the *anima mundi* are one and the same; that Moses learned magic from the Egyptians; and, finally, that Jesus Christ, too, was a magus. Any one of these would have ensured that Bruno be roasted alive – perhaps the Inquisition was furious Bruno had only one life to lose in the crackling flames.

MINERVA'S MAN

Bruno was sent to the pyre on 17 February 1600, ironically, or maybe deliberately, the day after Ash Wednesday, recalling the title of his infamous book. First he was taken from prison to the Inquisition's basilica, where he was handed over to the secular authorities (as was the procedure for the execution of heretics). Bruno may have seen the choice of location either as a cruel irony or perhaps as a source of comfort. The basilica is dedicated to Santa Maria sopra Minerva and was built on the foundations of a pagan Roman temple, which we now know was dedicated to Isis. However, when the basilica was built the deity was mistakenly identified as Minerva, the Roman goddess of wisdom and magic (among other things). Ironically, Minerva was the form of the goddess to whom Bruno had specifically chosen to dedicate himself.

From the basilica, Bruno was led to the Campo de' Fiori (Field of Flowers), then a meadow (and now a market square in the heart of the city), where he was tied to a stake and burnt alive. Schoppe says that he turned his head aside when offered a crucifix to kiss – demonstrating he was a pagan Hermeticist to the very last, and perhaps indicating that his alleged recantation was in fact an Inquisitorial invention. Or perhaps, in the one final moment when he had nothing to lose, he felt that he could reveal his true self.

In 1870, when the city of Rome passed from the control of the Pope to secular authorities, there were immediate calls to erect a statue in Bruno's honour in the Campo de' Fiori. Luminaries such as Herbert Spencer, Victor Hugo and Henrik Ibsen supported the petition. This is probably what prompted the Pope of that time, Pius IX, to call for the documents on Bruno's trial that were later found in his personal papers. However, it took until 1889 for the bronze statue, showing a rather sinister Bruno in his monk's robes and cowl, to be erected. The statue is today the focus for a variety of pilgrims even though they tend to be atheists, freethinkers and New Agers. But the original driving force behind the statue was Italian Freemasonry – the sculptor, Ettore Ferrari, was Grand Master of Italy and the statue was unveiled with the Campo 'festooned with flags bearing Masonic symbols'.[39]

The nineteenth-century adulation of Bruno was based on a serious misconception, which endured because of the gap in the official records. Many had come to believe that Bruno had been put to death solely for advocating the heliocentric theory or the infinity of worlds, making him a kind of forerunner of Galileo. This belief encouraged what one commentator calls 'a misguided interpretation of Bruno as a martyr for science'.[40]

Bruno was actually a martyr for Hermeticism. Although there was a connection with the Copernican theory, but Bruno was condemned not for preaching heliocentricity, but because of its special significance to him, particularly his vision that proving it would herald the coming Hermetic age.

Even today, the Catholic Church's attitude to Bruno remains startlingly unchanged. When, in the Holy Year of 2000, a suggestion was made that Pope John Paul II might finally forgive him – as they had Galileo – the official response was that Bruno 'had deviated too far from Christian doctrine to be granted Christian pardon'.[41]

But the question remains: why had it taken eight years for Bruno to be condemned? Why had his teachings suddenly become too hot for the Inquisition?

We suggest that the answer to these questions lies in events of a few months before, in an attempt to establish the Hermetic republic on Earth by force.

CHAPTER THREE

GALILEO AND THE CITY OF
THE SUN

Bruno's exit from Padua for his fateful stay with Zuan Mocenigo left a space on centre stage for others to move in. This certainly marked a major opportunity for one aspiring scholar. Bruno had applied for the then-vacant chair of mathematics at Padua University, but owing to his untimely arrest the job went to another candidate – none other than Galileo Galilei.[1] Of more immediate significance, however, was the arrival in Padua, just a few months after Bruno's departure, of a rising star of the Hermetic world who was his spiritual heir.

The similarities between the careers, philosophies and aims of Bruno and Tommaso Campanella (1568–1639) are so striking that they must have been working to the same plan. Indeed, twenty-three-year-old Campanella's arrival in the same circles so soon after Bruno's arrest suggests that he was picking up where the Neapolitan had been forced to leave off. And despite dramatic reversals of fortune, Campanella 'very nearly succeeded in bringing off the project of a magical reform within a Catholic framework, or, at least, in interesting a number of very important people in it'.[2]

Like Bruno, Campanella was born in the Kingdom of Naples, though much further south in the town of Stilo in the Calabria region, in 1568, which made him twenty years

Bruno's junior. Also like Bruno, and probably for the same reason of being a bright lad from humble origins – his father Geronimo was a cobbler – Campanella began his career in the Dominican Order, which he entered at the age of fourteen. After his novitiate he became a friar (a brother who lived in the outside world) rather than a monk like Bruno.

Campanella's own freethinking earned him the suspicion of heresy. In particular, he advocated that knowledge should come from the direct study of natural phenomena (remember the Hermetic motto: 'follow nature'), rather than from officially approved books. Not only was this – to modern eyes perfectly reasonable – approach deemed misguided but actually attributable to the Devil.

One of the major influences on Campanella's thinking was Marsilio Ficino, whose work was probably also responsible for attracting him to Hermeticism. Another esoteric influence was the venerable polymath Giovanni Battista della Porta (c.1535–1615), author of the classic 1558 treatise *Natural Magic* (*Magiae naturalis*), with whom Campanella struck up a friendship during a two-year stay in the city of Naples in the early 1590s. As with Bruno, Campanella was open to every sort of idea, but Hermeticism was the glue that held them all together and gave all human knowledge a recognizable shape.

Della Porta's influence inspired Campanella to write his first book, which advocated the practice of magic. Although it was only published in 1620, *On the Sense of Things and of Natural Magic* (*Del senso delle cose e della magia naturale*), argued that the world is a living thing and for the existence of the *anima mundi*. At around this time he also wrote *On Christian Monarchy* (*De monarcha Christianorum*), agitating for a reform of society and the Church. Clearly he was another Neapolitan destined to give the Vatican sleepless nights.

In 1592 Campanella travelled to Padua on the well-worn path via Rome and Florence, meeting Gian Vincenzo Pinelli and Padua University's new Professor of Mathematics, Galileo.[3] Campanella and Galileo were to stay in touch for the rest of their lives. It was also in Padua that more questions were raised about Campanella's dangerous beliefs. As a result, early in 1594 he was arrested by the Inquisition and transferred to Rome towards the end of the year – to the same prison as Bruno, although it is unlikely that they were allowed to communicate. Compared to Bruno's continuous imprisonment ending in his execution, Campanella got off lightly. After agreeing to abjure his works he was released into a kind of house arrest in a Dominican monastery, although in 1597 his superiors ordered him back to Naples. Campanella had not been around long enough to make himself as much of a nuisance as Bruno, and he had not so far made much headway with plans for Hermetic reform.

In fact, Campanella shared Bruno's vision of the great magical transformation that was glimmering over the horizon, and which was written in the stars. He also regarded the heliocentric theory as the trigger of the new age of Hermetic enlightenment, and – for astrological and other reasons – he believed it was destined to happen in 1600.

The approach of the new century encouraged Campanella to be much more politically proactive than Bruno even at the height of his career. Leaving Naples for the south, he threw himself into organizing the Calabrian revolt, which aimed to overthrow Spanish rule, beginning with Calabria – the arch of the Italy's 'foot' and 'toe', which had long been 'restive with political and religious dissidents'[4] – and then the whole of the Kingdom of Naples.

The Calabrian revolt is remarkable for the number of its Dominican supporters. Indeed, there was something very odd about the Order in Calabria, from at least the time it produced Bruno, but frustratingly after so many years it is impossible to pinpoint exactly the reason for this. This uprising was considerably more than just an expression of Calabrian nationalism. It was to be a preparation for the coming age, and aimed to establish a republic based on magical principles that would – under its messiah Campanella – hold aloft the torch of the new age for the rest of the world to follow. Bruno, too, had railed against Spanish rule over the Kingdom of Naples in *The Expulsion of the Triumphant Beast*.

If the revolt was successful it would bring the Hermetic republic geographically close to the Papal States – the two shared a long border, cutting across the whole of Italy from Mediterranean to Adriatic coasts. A truly alarming prospect for the Pope and his henchmen.

The uprising, however, was not to be. Informants betrayed it to the Spanish authorities, and after the organization was ruthlessly crushed in November 1599, Campanella and the other leaders were arrested. This almost certainly accounts for the Inquisition's sudden desire to be rid of Bruno, the revolt's spiritual inspiration, and he went to the stake barely three months later. Stephen Mason of Cambridge University argues that he was executed as an example to the Calabrian rebels, because of the connection to Campanella, and that he had been held for so long as a kind of hostage because of his standing among the insurgents.[5] Publicly executing their spiritual leader at the beginning of their special year – 1600 – would also have been a calculated psychological move, rather akin to roasting the Pope on 25 December of a new millennium.

This was, however, by no means the end of Campanella's story. His continuing career sheds a rare light on Galileo's trial thirty years later – over which Bruno, too, would cast a giant shadow.

Campanella escaped the death penalty visited on the revolt's other leaders through feigning madness. According to the law of the times, the insane could not be sentenced to death, not out of compassion but because they couldn't comprehend the opportunity to repent of their sins before execution. If a judge did condemn them he, not the condemned, would take responsibility for their eternal damnation. However, there was considerably more to feigning insanity than a bit of Hamletesque raving about clouds looking like camels and some foaming at the mouth. The madness defence was hardly the easiest option. To prevent every miserable prisoner from using it to evade the death penalty, the Neopolitan authorities had come up with a twist. The accused had to maintain their mad behaviour – or keep up the pretence – under prolonged torture.

Somehow the extraordinary Campanella managed to pass this test, and was duly sentenced to life imprisonment. For the next quarter of a century he was moved around a series of castle dungeons in the Kingdom of Naples. Although most prisoners in that place and time would have suffered horrors from the stark loneliness and the squalor of their own filth in the dark, fending off rats, Campanella's life was surprisingly non-onerous. Viewing his imprisonment as an extended opportunity for study and contemplation – much like being in a monastery – he spent his time refining his ideas and writing. Not only was he supplied with books and writing materials and had at least some light in his cell, but he also received a steady flow of scholarly visitors, mainly from Germany, who took his writings back home to be published. Why his jailers were so obliging is a bit of a puzzle, especially as it must have dawned on them by now that he was as sane as they were – probably more so. Presumably bribes were involved from somebody, somewhere.

The revolt having failed, Campanella's goal now became the reformation of society through the Vatican and, perhaps oddly, the Spanish monarchy he had plotted to overthrow. Like Bruno, his ambitions were nothing if not excessive.

Only once in his books did Campanella mention Bruno directly – significantly in a defence of Galileo published from prison in 1622 – and even then he was careful to declare that Bruno was a heretic. But Campanella was manifestly familiar with his philosophy and writings, judging by allusions in his work, his favourite being *The Ash Wednesday Supper*. Of course, given Bruno's fate and the continued opprobrium attached to his name, there was no way Campanella could be more open, especially given that he was trying to win support for Catholic reform – and doing so from prison.

Campanella's major work is *City of the Sun* (*Civitas Solis*), written in the first years of the 1600s but not published until 1623, in Frankfurt.[6] Basically concerned with a utopian society, the text takes the form of a dialogue between the Grand Master of the Knights Hospitaller and the captain of a ship that had sailed to the New World. The captain relates how, after being shipwrecked, he was found by the inhabitants of the City of the Sun, describing its society in detail to the Grand Master. Clearly Campanella's ideal republic, the kind he had hoped to establish in Calabria, the City of the Sun is designed and run according to magical and astrological principles. It is a Hermetic-Egyptian utopia, derived from the prediction at the end of *Asclepius'* Lament. George Lechner of the University of Hartford, a specialist on magical and astrological symbolism in Renaissance art says of *City of the Sun*: 'In it, Campanella developed the notion of a new city-state, led by a philosopher-priest-king, and guided by Hermetic magical principles.'[7] And of course it is no coincidence that it was a city of the *sun* that was being debated, reminiscent of the '*Civitas solis*' that Bruno discussed with the librarian of the Abbey of St Victor in Paris, saying that the 'Duke of Florence' planned to build it.[8]

Even from prison Campanella played an influential role in events surrounding the next great champion of the sun-centred theory: Galileo Galilei. The Hermetic chain remained unbroken.

THE THRICE-GREAT TRIO

Giordano Bruno had made heliocentricity the centre of his Hermetic revolution, the sign that would trigger either the downfall or the reformation of the Church, neither of which was regarded with any great enthusiasm by the Vatican. For Bruno and the Giordanisti, heliocentricity was not just a theory: they believed its acceptance would usher in a new Hermetic utopia. And even with Bruno out of the way, it was feared that he had left behind a secret society – who and where nobody knew – which was proactively com-mitted to bringing the Hermetic revolution about. Tommaso Campanella, Bruno's spiritual heir, who shared his view of the importance of heliocentricity and was possibly even one of the Giordanisti, had conspired in a rebellion against the Kingdom of Naples and therefore against the Spanish crown, aiming to attack those who were deemed most loyal to the Catholic cause.

Given this context, Copernicus' original evocation of Hermes Trismegistus' name in *On the Revolutions of the Celestial Spheres* was hardly likely to have been missed by those whose job it was to protect the Church. Perhaps placing the sun at the centre had been a devilish Hermetic plot all along? There was no way for those organizations whose task it was to defend the Church – the Inquisition and the Jesuits – to be sure, and every reason for them to be nervous. During the sixteenth century the Roman Church had only just survived its greatest trauma, a seemingly impossible undermining of its authority by the rise of the Protestant Churches. So who was to say what might happen next? The ideas of Bruno and other Hermeticists were being discussed across Europe, and even highly placed members of the Catholic Church had embraced them. Hermetic principles were being openly advocated. And then there were the Giordanisti – how many there were, and how widely they were spread, nobody knew. Maybe the Inquisition and Jesuits were over-reacting, but these were times that engendered paranoia. And so it was considered that – at the very least – establishing heliocentricity would attract more converts to Hermeticism. More readers would devour Bruno's works, and possibly attempt to act on his agenda of radical reform.

As long as Copernicus' idea remained simply a theory, however, the Hermetic implications barely registered. But when an individual claimed he had come up with *proof*, then the Church began to become seriously worried. And ecclesiastical anxiety ran even deeper when it was discovered that the threat came from a direct associate of the mystical revolutionary Tommaso Campanella and other Giordanisti suspects, such as Pinelli and his circle in Padua – in other words, Galileo.

The Hermetic interpretation of heliocentricity adds an important and otherwise missing element to the story of Galileo's persecution, finally making sense of some of its more puzzling aspects. Why, for example, were the Jesuits – Galileo's main enemies – so zealous about making an example of him? And why exactly did they consider his work so dangerous?

Galileo wrote to a friend in Paris as he was about to leave for Rome to face the Inquisition in 1633:

I hear from a good source that the Jesuit Fathers have impressed the most important persons 'in Rome' with the idea that my book 'the Dialogo' is execrable and more dangerous to the Holy Church than the writings of Luther and Calvin.[9]

Comparing Galileo's work to Luther and Calvin seems rather excessive. How could proving Copernicanism possibly do anything like the same damage to the Church as those famous pioneering Protestants? And during a time when other heretics were challenging fundamental doctrines such as transubstantiation, heliocentricity does seem rather tame. There was something else behind the Church's anxiety, something massive but unstated which lies somewhere in the significance of the heliocentric theory to the dangerous Hermeticists.

Because the Galileo affair has been used for so long to score points in the contest between science and religion it has become hedged round with assorted myths propounded by one side or the other. Take for example the well-worn story of Galileo finishing his public recantation of his belief in the motion of the Earth around the sun by muttering the aside, 'And yet it moves'. This was invented a century after the event, but has been repeated so often it is now considered by many to be the gospel truth. With so many assumptions and so many myths, it is almost impossible to uncover the simple truth. Almost, but not quite.

Galileo has often been depicted as a modern rationalist-materialist scientist who had somehow been born out of time, and who was persecuted by superstitious – in other words cretinous – men whose intellects were stuck in the Middle Ages. Galileo is seen as a martyr for science and a victim of irrational religion. But of course the reality is that he was very much a man of his time, and we should no more assess his character and motivation by modern standards than we should Copernicus or Kepler.

While most educated people today still think that

Galileo's trial was all about a clash between the scientific and religious mindsets, historians have long realized that this is way off the mark. It has therefore become fashionable to see the affair as a collision between two great and obstinate egos, two pathologically 'right men': Galileo, who refused to be told what he could do or say, and Pope Urban VIII, whose ego had been bruised by Galileo putting his views in the mouth of a character offensively named Simplicio. The prevalent view is that if only Galileo had not been so stubborn, and had made it clear that he was presenting heliocentricity simply as a hypothesis, then all of his trauma could have been avoided. The very fact that the myth of the clash of egos has endured is an acknowledgement that something is still missing. It seems that the elusive 'something' may have been a factor that neither side wanted to see the light of day . . .

On the question of Galileo's attitude to Hermeticism, ironically other historians argue that he would have nothing to do with it because he was too staunch and conventional a Christian. Particularly after the way he was portrayed in Dan Brown's thriller *Angels and Demons* (2000) there was a rush to paint him as an especially devout Catholic, respectful of the Church. But there is little evidence for this. Galileo's published works deal with matters of science, not religion, and his surviving personal letters contain very little on religious matters. Naturally he used the conventional Christian platitudes of the time, and observed the outward trappings – going to church, taking communion and so on – as everybody was compelled to do in that time and place; but no more than this.

In his published works, Galileo explicitly distanced himself from certain of the esoteric arts (most specifically numerology derived from Pythagoras), which is taken by today's commentators to indicate his modernity and rationalism. However, given what had happened to Bruno, this could equally have been simply an act of self-preservation: one specialist, Giorgio de Santillana, specifically links the disavowal of numerology to Galileo distancing himself from Bruno and his ilk.[10] And in any case, dismissing one arcane system does not necessarily mean dismissing everything esoteric. And yet on the other hand, Galileo practised astrology. It is often stated in popular histories that, although he drew up horoscopes for wealthy clients, he only did this for the money, and never actually believed in it. In fact, there's no evidence at all that this was his attitude – it is yet another example of modern projection.

Galileo was undoubtedly a brilliant pioneering scientist who used observation and experiment to work out the laws governing physical phenomena and sought to explain them in mathematical terms. The methods he developed would inspire and shape the next generation and culminate in the genius of Isaac Newton. Both Einstein and Stephen Hawking have hailed Galileo as the father of modern science, and he has been described as 'the world's first celebrity scientist'[11] – the Einstein of his day. But there are many ironies in his story and the way it has passed into history, or perhaps more precisely, legend.

The first irony is that what Galileo is best known for now – helping to establish the heliocentric theory – is actually one of the least important aspects of his work. His major contributions to science were in what we today would call the field of physics: motion, optics, acoustics and so on. In astronomy, his big innovation was to improve the telescope to the point that it was good enough for astronomical observations (although he originally thought in terms of military and maritime applications). And while the observations Galileo made with the telescope produced new evidence in favour of Copernicus, the arguments he thought proved the theory were, in fact, entirely mistaken.

Galileo thought that the smoking gun was the phenomenon of the tides, arguing their ebb and flow could only be explained by the Earth's rotation, airily dismissing Kepler's suggestion that they were caused by the pull of the Moon. In this, Galileo was, of course, completely wrong.

In fact, his whole attitude to heliocentricity was at odds with the methodical and meticulously worked-out approach that characterized the rest of his work and which rightly justifies his status as the founding father of the modern scientific method. Einstein thought Galileo was so determined to prove Copernicus right that he was blind to the obvious problems with his argument.[12] As the Danish science historian Olaf Pedersen, speaking at a conference on the Galileo affair in Cracow in 1984, observed:

> In consequence [of his acceptance of the theory] it became imperative to find convincing reasons for its being true in a physical sense, as Galileo tried to do with his somewhat unsatisfactory theory of the tides . . . [13]

In other words, Galileo became convinced by the theory and then set out to find evidence for it – hardly a true scientific approach. He enjoyed his celebrity status and the material benefits it brought. He had a flair for self-publicity, never being one to hide his innovations and discoveries, if anything exaggerating them. But he seems to have made it his mission in life to see the theory of heliocentricity proved, while being uncharacteristically circumspect about his support for it. Although writing to Kepler as early as 1597 that he had 'become convinced by Copernicus many years ago',[14] publicly he was keen to be seen as much more equivocal, even evasive.

Of course, Bruno's fate must always have been at the forefront of his mind, and must have acted as a hideous, cautionary tale. Advocating the motion of the Earth had certainly contributed to his condemnation as a heretic, and Galileo, along with other scholars in Catholic lands, may well have considered it prudent not to whip up any hype. But despite this, there is evidence of a more solid connection between Galileo and Bruno and the Giordanisti – including evidence that Galileo owed an intellectual debt to Bruno. And there is no doubt whatsoever that Galileo was fully aware of the significance that the Hermeticists read into heliocentricity.

Galileo was a lifelong friend of Campanella. One of his staunchest supporters during the controversy, Campanella composed the *Defence of Galileo* from his prison cell in 1622. And ten years later, by then a free man living in Rome under the protection of the Pope himself, he was still corresponding with Galileo during the latter's most difficult time, urging him to stand firm *because of the spiritual importance of his work*. Yates remarks when discussing *Defence of Galileo*:

> Campanella is being careful to dissociate himself from the full implications of Bruno's Copernicanism. This was all the more necessary since, both in the apology and in letters to Galileo, Campanella speaks of heliocentricity as a return to ancient truth and as portending a new age, using language strongly reminiscent of Bruno in the *Cena de le ceneri* [*The Ash Wednesday Supper*] . . . And in other letters he assures Galileo that he is constructing a new theology which will vindicate him. It has therefore to be made clear that heliocentricity as a portent of a new age, and as integrated into a new theology did not mean for Campanella at this stage in his career, acceptance of all Bruno's heresies.[15]

So Galileo was not only in contact with Hermeticists, but was also very aware of just how important they considered his work. But could the connection go much deeper? Was there a more mystical dimension to the whole affair?

Galileo was familiar with Bruno's writings. In the 1590s, when he first focused on heliocentricity, there was no problem with being a fan of the Neapolitan – just as after 1600 there were excellent reasons not to be seen to be. After the publication of Galileo's first book touching on the controversy, Kepler criticized him for not honestly acknowledging the intellectual debt he owed to Bruno.[16] Of course it was easy for Kepler, who cited Bruno in his own work, to criticize Galileo from the safety of Bohemia.

But Galileo's interest in Bruno goes deeper than merely reading his books. There are close parallels between Galileo's 1632 *Dialogue Concerning the Two Chief World Systems* – which led to his downfall – and Bruno's *The Ash Wednesday Supper*, the first of his works to advocate Copernicus and to declare that establishing heliocentricity would free the human spirit. It may not be a coincidence that this was Campanella's favourite of Bruno's works.

Another clue suggesting Galileo's familiarity with Bruno comes from a passage in the *Dialogue* where he lays the foundation for the later theory of relativity. Although the term is popularly associated with Einstein, what he formulated were his special and general theories of relativity, which are in fact highly complex developments of Galileo's original principle (sometimes called 'Galilean relativity'). This argues that physical phenomena can only be properly described according to the context in which they are observed – i.e., the same event can look completely different to observers in different places. This principle underpinned Newton's laws of motion and Einstein's own theories.

In *The Ash Wednesday Supper*, published over forty years

earlier, Bruno made the same point with a very similar example: if two people, one on shore and the other on the deck of a moving ship, drop a stone, each will see their own stone move through an identical path, dropping the same distance at the same speed, but they will perceive the other's stone as moving further – not only downwards but sideways – and therefore faster, since it covers a greater distance in the same time.[17] Descriptions of events therefore depend on the frame of reference.

Having never publicly referred to the Hermetic interest in heliocentricity, why should Galileo base his masterwork on a book by someone anathematized by the Church for championing precisely that theory? Perhaps this was a covert acknowledgement of his debt to Bruno, or even a coded hint that he was aware of his own significance to the Hermetic vision.

THE DAY THE EARTH STOOD STILL

Galileo di Vincenzo Bonaiuti de' Galilei's career began in 1592 at the age of twenty-eight, when thanks to Bruno's incarceration he became Professor of Mathematics at the University of Padua. This is where he met Campanella and began an important and lifelong association. Another major influence at that time was Pinelli – often described as Galileo's mentor – who introduced him to the emergent science of optics, which was to make Galileo's reputation. Another of his dubious associates was Traiano Boccalini, author of the Bruno-inspired *News from Parnassus,* and a controversial friar and professor of canon law named Paolo Sarpi, who was at the forefront of the legal challenges to the Pope's authority and the attempts to forge an alliance with James I's England in the first decade of the seventeenth century. With friends like these, the Inquisition must surely have kept a very close eye on Galileo from the beginning.

Galileo became convinced of the truth of the Copernican theory 'many years' before 1597, although precisely why he had this epiphany remains uncertain. We have also seen that he incorrectly considered the movement of the tides as the best evidence for, even the proof of, the theory. He persisted in this view even when he produced much better evidence through his pioneering use of the new cutting edge technology of the telescope, begun around 1610. His astronomical observations – that the Moon's rugged surface is reminiscent of our own world, the existence of the moons of Jupiter and particularly Venus' lunar-like phases – strongly supported Copernicus' theory. Galileo realized how sensational these discoveries would appear, and cannily sought to use them as leverage to build a career. So he rushed into print before anyone could steal his thunder, premiering his first wave of discoveries in *Starry Messenger* (*Sidereus nuncius*) in 1610.

As he had guessed, the intelligentsia became greatly excited and he landed the position he craved as court mathematician and philosopher to Cosimo II de' Medici, Grand Duke of Tuscany. Perhaps this wasn't too surprising given that Galileo had been careful to dedicate the book to him and proposed calling the new moons of Jupiter the 'Medicean stars'. Even the world's loftiest thinkers obviously recognized the most basic principle: flattery will get you anywhere.

It seems odd that Galileo failed to use his discoveries to bolster the Copernican theory, even though he was an ardent supporter. In both *Starry Messenger* and a follow-up book on his discovery of the phases of Venus, he merely presented the observations. Perhaps, as he was hoping to build a glittering new career on them, he decided that it was best to play down the Copernican implications of his discoveries.

But the row refused to go away. Most readers with an

astronomical background got the point: Galileo's discoveries seriously undermined the traditional Ptolemaic system. But even this failed to shift the consensus to Copernicus. Hybrid systems, such as Tycho Brahe's, where some celestial bodies orbited the sun and some the Earth, were preferred.

From the Church's point of view Galileo's discoveries were already unwelcome news, and threatened worse to come. Not only was his work propelling scholars towards heliocentricity, but the telescope might lead to further discoveries that would decisively tip the balance in its favour. And now there was an added piquancy: if irrefutable proof was forthcoming, would it inspire the Hermeticists to kickstart their revolution, philosophically, theologically – even politically?

Matters came to a head in 1615 when Galileo finally went public with his support for heliocentricity. He circulated an essay based on the biblical passages that implied the Earth did not move, including the unequivocal statement: 'I hold that the Sun is located at the centre of the revolution of the heavenly orbs and does not change place, and that the Earth rotates on itself and moves around it.'[18] This was an extraordinarily dangerous declaration that would transform Galileo's fame into notoriety overnight.

Pope Paul V ordered a group of cardinals to investigate the issue of heliocentricity on theological grounds, and they decided it was contrary to scripture. As a result, Copernicus' On the Revolutions of the Celestial Spheres was finally banned, along with any other pro-heliocentric works. Galileo was summoned to Rome to be warned off and put right. The sun moved round the Earth and not vice versa. It was true because the Vatican said so.

But there was an unspoken subtext: the cardinal tasked with warning Galileo was none other than Roberto Bellarmino, the same man who had interrogated Bruno in his last months, and was responsible for his condemnation and execution. This was not a coincidence – Bellarmino had been Archbishop of Capua since 1602, but was recalled to Rome specifically to deal with Galileo. He even interviewed Galileo in the same room as he had interviewed Bruno.

Bellarmino, of course, understood from his experience of Bruno the significance that heliocentricity possessed for the Hermetic revolution. Bruno was dead and Campanella incarcerated in Naples, but they had followers – nobody knew how many. And now here was Galileo, associated with both Campanella and Pinelli, getting dangerously close to the proof that Bruno had declared would trigger the new Hermetic age. In the end, nothing harsh was done to Galileo. He was simply given a document written by Bellarmino himself stating that the Pope had decreed that Copernicus' views could not be 'defended or upheld'. Galileo hastily agreed.

Even more telling is Galileo's immediate reaction after receiving his warning. Rather than return directly to Florence, he wanted to travel to Naples and was obliged to request permission from his patron, Duke Cosimo – but Cosimo refused. Why Naples? A crucial piece of the jigsaw fell into place when we read in a paper by Olaf Pedersen, a specialist in the religious aspects of the Galileo affair, that the reason for Galileo's request and the odd refusal was that he wanted to visit Tommaso Campanella in his prison cell.[19] In other words, the Church brings in the man who had condemned Bruno to warn Galileo off, and Galileo wants to consult Bruno's successor Campanella; surely none of this was a coincidence.

Having been denied a meeting with Galileo, Campanella rallied to the cause, penning the *Defence of Galileo*, which his followers published in Frankfurt. However, given Campanella's reputation – one conviction for heresy and another for subversion, for which he was still doing time – the kind of support he could muster was hardly designed to enhance Galileo's reputation. Which is probably why, back in Florence, Galileo kept his head down. Nothing in the Pope's decree prevented the *discussion* of heliocentricity as a hypothesis, and many scholars were avidly doing just that. However, Galileo himself dropped the whole subject for many years, although he was clearly waiting for an appropriate time to re-emerge as its iconic figurehead.

A potential change for the better came in 1623 when one of Galileo's old friends, Maffeo Barberini, became Pope Urban VIII. They had met at the Florentine court, and Barberini was an admirer of Galileo's work, especially his research into the laws of motion. Galileo went to visit Urban in Rome the year after he was elected, and they had six private meetings – during which, as he himself reported in a letter to a friend, Galileo described all believers in Copernicus' work as 'heretics'.[20] Clearly he had no desire for another confrontation with a Bellarmino clone.

In another of those astonishing reversals of fortune that litter the history of that era, Urban's election was also good news for Campanella. In 1626 Urban requested that the Spanish king release him from prison so he could travel to Rome to perform protective magic to ward off the evil effects of an eclipse that the Pope's enemies had predicted would kill him. After twenty-seven years, not only was Campanella free but appointed adviser to the Pope. Urban even went so far as to grant him permission to found a college in Rome to train missionaries who espoused his religious and philosophical ideas. Such papal favour being bestowed on his greatest and most controversial supporter was another good sign for Galileo. In 1631, the year before it all fell apart, Urban even appointed him as a canon, which enabled him to draw income from two vacant benefices (without doing a day's work in either).

It was during this time that Galileo decided it was safe to have another stab at pushing the heliocentric theory. And so he wrote *Dialogue Concerning the Two Chief World Systems* – unusually for him in Italian rather than Latin, widening his potential readership – in which two scholars debate the Copernican and Ptolemaic systems, with a third adjudicating. It was published in Florence in 1632, having been granted formal approval by the Inquisition in that city. Galileo had even sought permission from Urban to publish; the latter only asked that his own views on the matter be included.

The irony – which is seldom mentioned by modern historians of science – is that the main pro-Copernicus argument that Galileo puts forward in the *Dialogue*, his old 'proof' based on the tides, was wrong. His original title was, in fact, *Dialogue on the Ebb and Flow of the Sea*. The Inquisition in Florence forced him to change the title, which is odd, as the new one made it more obvious that the book was about the heliocentric debate. Galileo was careful to keep to the rule of discussing Copernicanism without actually advocating it. Nevertheless, the book caused rumblings, especially among the Jesuits, and Urban came under pressure to act.

Despite the myth of the 'clash of egos', it is clear that Urban had to be pushed into action. His position as pope was far from secure, as many in Rome thought him too soft on Protestantism – there was even talk of deposing him.[21] This was largely because Urban was concerned about the power of the Hapsburg dynasty, which ruled Spain and the Holy Roman Empire, both of which were locked in battle with the Protestant nations. For his own political reasons he had refused to give his sanction to the war or to lend it diplomatic or military support, but it did lead some to wonder where his sympathies really lay. His many opponents among the Cardinal Inquisitors were making much of his endorsement of the *Dialogue*'s publication as another sign of his softness on heresy. He therefore had to take action to keep his own position secure. This was no clash of egos. Urban was just running scared.

As a result of Jesuit pressure, Urban appointed a commission to investigate whether Galileo had broken his ban of sixteen years earlier. Some historians believe that this was an attempt to keep the Inquisition out of the matter, another sign of the Pope's reluctance to let the Inquisition loose on his old friend. If so, it was remarkably unsuccessful. In September 1632 Urban instructed the Inquisition in Florence to deliver a summons to a shocked Galileo to present himself in Rome to answer questions about his book. He appeared before the Inquisition in April the following year, no doubt with Campanella's advice to stand firm – because of the *theological* (that is, Hermetic) importance of establishing that the sun was at the centre – ringing in his ears.

Galileo's defence was that his book had not upheld Copernican theory, but had merely discussed it. He declared that until the decree of 1616 he had regarded neither the Copernican nor Ptolemaic hypothesis as beyond dispute (contradicting his statements to Kepler thirty-six years earlier), but since then he had held the Ptolemaic view 'to be true and indisputable'.[22] While few would blame Galileo for reneging on his own opinions and weaselling out of the situation – after all, this was the Inquisition he was facing – these were hardly the words either of a noble defender of intellectual freedom or willing would-be martyr. And yet neither does he seem an arrogant old man who refused to admit he was wrong.

Galileo lost. The inquisitors decided that the *Dialogue* was a disingenuous attempt to promote heliocentricity, which it probably was, and that his attempts to disguise it as a mere discussion were totally unconvincing. He was found *'veementemente sospetto d'eresia'* – vehemently suspect of heresy – just one degree below actually being a heretic. The only way out was to 'abjure, curse and detest' the very ideas that caused the suspicion.

Galileo had to admit his error and renounce his ideas, kneeling before the altar of Santa Maria sopra Minerva, the same basilica from which Bruno had set out to his horrendous death thirty-three years earlier. Publication of anything by Galileo – anything he had written or would write in the future – was forbidden (although in the event he did manage to get some works printed in Germany). He was sentenced to life imprisonment, but as he was over seventy years old, was instead committed to house arrest. He stayed first with a supporter, the Archbishop of Siena, where one of his first visitors was none other than Tomasso Campanella . . . [23]

Later, Galileo was allowed to return to his own villa outside Florence, where he died in 1642. Less than a year before his death he wrote to the Florentine ambassador in Venice that:

> The falsity of the Copernican system ought not to be doubted in any way, and most of all not by us Catholics who have the undeniable authority of Holy Scripture, interpreted by the best theologians.[24]

Perhaps Galileo had an unusually over-developed sense of irony.

But what of Campanella? In 1634, the year after Galileo's trial, there was another attempt to organize a revolt in Calabria. Whether Campanella was directly involved is unclear, but the leader was certainly one of his followers. So it was expedient, to say the least, for him to leave Rome for Paris – a well-worn route for fugitive Italian Hermeticists. There he became a favourite of Cardinal Richelieu, who persuaded the king to give him a pension. Encouraged by this, he transferred his hopes to the French monarchy, urging Richelieu to make Paris into his City of the Sun. His big hope settled on the future Louis XIV, born in 1638, who he expected to rule the world in partnership with a reformed papacy. Campanella was the first person to call the infant Louis the Sun King, as an acknowledgement of his great *Hermetic* potential.[25]

After Campanella's dizzyingly strange and extreme career, which took him from castle dungeons to the favour of some of the greatest figures in Europe, he died in Paris in May 1639. But there can be no doubt that his legacy lived on.

GALILEO'S SECRET

Although during the nineteenth and first half of the twentieth centuries researchers perceived a connection between the trials of Bruno and Galileo, the notion of Bruno's fate being a more severe foreshadowing of Galileo's persecution, dying for his Copernican beliefs, is not borne out by the facts. There was indeed a connection between the two, but it is the other way round. Action was taken against Galileo because of the Hermetic – the Brunian – implications of his views.

Yet while not often recognized, the connection between the two trials is hugely significant. Although Galileo's trial is always cited as *the* moment when forces of reason and dogma collided head-on, the Hermetic factor is arguably the most important. It was, after all, the reverence that heliocentricity was accorded by Hermeticists in general and Bruno's followers in particular that was the major reason the Church sought to damn heliocentricity, and therefore Galileo himself.

Neither side could admit what Galileo's trial was really about. While being aware of the Hermetic implications of the *Dialogue*, Galileo never made them overt, which meant that the Church couldn't use that against him. It is unlikely it would have wanted to draw attention to the importance of heliocentricity for the Hermetic revolution in any case. The Hermetic factor was therefore present, however, but simply relegated to the background – which is why there is a distinct sense of something missing in the conventional story of the trial.

Given the uncompromising Bruno and the revolutionary Campanella, the Inquisition and the Jesuits would have undoubtedly been only too fearful of the threat posed by Hermeticism. They would have traced the same connections we have outlined – beginning with Copernicus' references to Hermes Trismegistus, through Bruno's reforming career and the hidden presence of the Giordanisti, to Galileo's links with Pinelli and, most damningly, Campanella. They may even have seen the connection between Galileo's *Dialogue* and Bruno's *The Ash Wednesday Supper*. Even if they were putting two and two together and coming up with five – a not uncommon occurrence with the Inquisition – these connections would still have shaped their fears and consequently their actions.

It seems, however, Galileo was by no means as innocent as he tried to appear. There are valid questions, for example, about his relationship to the secret Hermetic reform movement. There is his continued association and correspondence with Campanella to take into consideration, especially his wish to see him in the wake of his warning-off in 1616. What would Galileo get out of such an association? Campanella was a religious, esoteric and political theorist – not a mathematician or scientist. For an ambitious man like Galileo, conscious of his image, Campanella was hardly the kind of company he should have wanted to keep.

And then there is Galileo's apparent use of Bruno's *The Ash Wednesday Supper* – which contains the first mention of the concept of the Copernican sun as the trigger for a new Hermetic age – as a model for his *Dialogue Concerning the*

Two Chief World Systems. Was this merely a belated, and necessarily covert, acknowledgement of Galileo's intellectual debt to Bruno, rectifying the failure for which Kepler had criticized him? Or was it a covert signal to the Giordanisti that he was a sympathizer – perhaps even one of them? It is safe to say that at the very least Bruno's work influenced Galileo's, which yet again places Hermeticism at the centre of the scientific revolution.

CHAPTER FOUR

THE FALSE ROSICRUCIAN DAWN

The Hermetic cause suffered several major setbacks in the early years of the seventeenth century, and for a time it must have seemed as if its hopes for a new golden age had been dashed once and for all. The first setback was, of course, the grisly execution of audacious prime mover Giordano Bruno in 1600, but the second came fourteen years later and was to provide even more ammunition for those opposing the Hermetic movement.

When the *Corpus Hermeticum* was rediscovered in the mid-fifteenth century everybody – whether they supported or opposed Hermeticism – accepted that the texts dated from the most ancient days of the Egyptian civilization. But suddenly a learned work exploded onto the scene that made the startling claim that the texts were of a much later provenance, not being written until the second or third century CE. The bombshell was *Of Things Holy and Ecclesiastical (De rebus sacris et ecclesiasticis)*, by one Isaac Casaubon. Born to refugee Huguenot parents in Geneva in 1559, he was widely regarded as the most learned man in Europe, his speciality being classical languages. After a glittering academic career in Switzerland and France he found himself working at the royal library in Paris under the patronage of Henri IV, the great hope of the Hermetic reformers. In May 1610 Henri, like his predecessor, was assassinated by a Catholic fanatic. This prompted a lurch towards ultra-orthodox Catholicism in France, which made life decidedly uncomfortable for Protestants such as Casaubon, who was more than happy to accept an invitation from James I to move to England.

Upon his arrival, the King of Scotland and England asked Casaubon to work on a rebuttal of a key text of the Counter Reformation, the gargantuan multi-volume *Ecclesiatical Annals* (*Annales Ecclesiasti*) by the Catholic cardinal Caesar Baronius – a sweeping history of Christianity that set out the historical case for the primacy of the Catholic Church and the correctness of its teachings. Unsurprisingly it articulated the conventional view accepted by many Catholic theologians that Hermes Trismegistus was one of the pagan prophets of the coming of Christ.

Casaubon only managed to write the first of many intended volumes giving a point-by-point critique of Baronius, as he died in July 1614, and was buried in Westminster Abbey. But in that single volume he still managed to deal Hermeticism a blow that to some seemed terminal, although ironically he had intended to demolish the *Christian* tradition that accorded Hermes Trismegistus a privileged place in pagan history, rather than to attack Hermeticism itself.

Casaubon began with the observation that no ancient author – nobody, in fact, before early Christians such as Lactantius and Augustine – even so much as mentioned Hermes Trismegistus, still less cited him as the fount of all wisdom. Intrigued by this, Casaubon compared the Hermetic texts with other works to try to establish their sources. He concluded that, contrary to tradition, the writers of the Hermetica had drawn upon the works of Plato and the books of Old and New Testaments. He argued, for example, that the sections of the *Pimander* that had once been thought to prefigure the opening chapter of John's Gospel were themselves really based on it.

Since most European readers had used Marsilio Ficino's Latin translation, Casaubon revisited the original Greek to analyse the language, using a printed edition that had been published in 1554. His heavily annotated copy is now in the British Museum. Discovering that the Hermetica's Greek dated from the early centuries CE rather than from antiquity, not unnaturally he concluded that the Hermetic texts were forgeries, created early in the Christian era in order to convert pagans to Christianity by building a bridge between their respective beliefs – a kind of ecclesiastical white lie. He accepted that although there had been a real person known as Thrice-Great Hermes in the high civilisation of ancient Egypt, the Hermetica was falsely attributed to him.

The implications for Hermeticists, particularly those who followed Bruno's extreme interpretation that Hermeticism represented the true original religion, were devastating. Their sacred books did not represent the wisdom of the ancient days of Egypt that produced the pyramids and the Great Sphinx after all. Their sacred books were no longer sacred.

For the historian Garth Fowden, Casaubon's work is 'the watershed between Renaissance occultism and the scientific rationalism of the new age'.[1] Yates called the medieval and Renaissance belief in the antiquity of the *Corpus Hermeticum* 'the great Egyptian illusion'.[2] Ironically, of all illusions, this had been remarkably productive – after all, it had created the Renaissance – but it was an illusion nevertheless.

The great disillusionment, however, was not an overnight sensation. It took a while for Casaubon's arguments to filter through, especially as they were buried in an otherwise obscure and scholarly critique of Baronius. Tommaso Campanella, for example, who continued his campaign for a Hermetic reform for another quarter of a century after *De rebus* was published, was either unaware of it, or rejected its message. And with a huge irony, it also failed to galvanize Catholic Europe. If the Church's scholars even bothered to read Casaubon, they preferred to side with Baronius and retain their traditional view of Hermes. As we will see, it took ten years for Casaubon's discovery to be used against Hermeticists, and a full half a century to become widely known and accepted.

Despite this blow, Hermeticists often argued that if the philosophy worked, its age and provenance were pretty much irrelevant. Particularly in England, some argued that while the texts themselves might be later than had been thought, the philosophy and cosmology that they contained were much more ancient, having been passed down throughout the centuries before being committed to writing. Perhaps along the way they had absorbed ideas from other philosophies, such as Plato's, but they still retained the essential beliefs of the Egyptians – a reconstruction that fits perfectly with recent findings. In fact, there were some glaring flaws in Casaubon's line of argument, which were recognized in his day and have become more apparent with the passage of time. Although we will deal with this more fully in a later chapter, suffice it to say here that modern discoveries show that Egyptian thinking was indisputably a major influence on the Hermetica. In addition, Casaubon's key argument that New Testament books such as John's Gospel had a direct influence on the Hermetica was refuted long ago. Whatever the Hermetic texts are, they are emphatically not Christian forgeries.

What was lost as a result of Casaubon's book, though, was the underlying belief, whipped up by Bruno, that the great reform would mark a return to the most ancient religion of all, the *prisca theologia*. Even so, the zeal to reform did not simply disappear, instead it found a new mode of expression. Indeed, in the years immediately following the execution of Bruno and incarceration of Campanella, the reforming spirit was already being repackaged with the aid of another major 1614 publication. And this was to cause high anxiety and even paranoia among Catholics for many years, and is still the subject of many conspiracy theories, hotly debated to this day.

'EUROPE IS WITH CHILD'

The second book of 1614 had a much more immediate impact than Casaubon's *De rebus*, one that has never really faded away. This was the appearance of the first of what became known as the 'Rosicrucian manifestos', which represented a key development of the reforming side of the Hermetic and esoteric tradition and launched a new and enduringly evocative term. The first of the two manifestos was *Fama Fraternitatis* (*Fame of the Fraternity*), or, *Discovery of the Order of the Rosicrucians* (*Fama Fraternitatis, dess Löblichen Orden des Rosenkreutzes*), usually known simply as the *Fama*. Written in German, it was published in Hesse-Cassel in Germany, but according to contemporary references had been circulating in manuscript for at least four years prior to being printed.

If ever a book caused a sensation among German philosophical circles, this was it. But the furore had barely subsided when just a year later its sequel appeared. *Confession of the Fraternity R.C. to the Learned of Europe* (*Confessio Fraternitas R.C., ad eruditos Europae*) – usually referred as the *Confessio* – was this time written in Latin and was clearly aimed at a more scholarly audience.

The manifestos announced the existence of a secret order, the Fraternity of the Rose Cross, and invited those who shared its ideals and aims to join. The *Fama* momentously declared that 'Europe is with child', trembling on the brink of a golden age. Great discoveries by recent generations had expanded mankind's knowledge of the world, the universe and nature, and had also ushered in a new appreciation of the magnificence and potential of humankind. In the words of the 1652 English translation:

> [God] hath raised men, imbued with great wisdom, who might partly renew and reduce all arts (in this our age spotted and imperfect) to perfection; so that finally man might thereby understand his own nobleness and worth, and why he is called *Microcosmus*, and how far his knowledge extendeth into Nature.[3]

This could have been Pico della Mirandola speaking, 130 years before.

The manifestos, however, went on to warn that the forces of popery and a rigid and outmoded scholarship were obstacles strewn in the path of the coming age.

Tantalizingly, the manifestos named no author, although the writer of a third work two years later, clearly continuing the theme, did eventually identify himself. This was entitled *The Chemical Wedding of Christian Rosenkreutz in the Year 1459* (*Chymische Hochzeit Christiani Rosencreutz anno 1459*). Although published anonymously, a Lutheran cleric and writer, Johann Valentin Andreae (1586–1654), claimed authorship in his autobiography. As he was a prolific writer of plays, allegorical stories and theological and philo- sophical essays, and the *Chemical Wedding* is clearly in his style, he was probably telling the truth. So was he also responsible for the manifestos?

Andreae certainly had a connection with the *Fama*, and almost certainly wrote at least parts of the *Confessio* while studying theology at Tübingen University. But opinions are divided about whether they are solely his works or, as is more likely, whether others were involved as well. It seems that the physician and esotericist Tobias Hess, Andreae's close friend and mentor, provided considerable input.

Perhaps the whole idea was his. Hess died in 1614, which would explain why the *Chemical Wedding* was a solo effort executed by the younger Andreae.[4]

The books outlined the foundation myth of the fraternity, which was, it claimed, created by 'C. R.' – Christian Rosenkreutz – who was supposedly born in 1378. He aimed to effect a mighty reform of the arts, sciences and religion, and intended to fix all the 'faults of the Church'. One can safely guess that such a man and such a shadowy organization, would hardly have been music to Vatican ears. Suddenly every shadow posed a potential Rosicrucian threat, every printing press a potential bombshell.

Interestingly, the *Fama* attributes Rosenkreutz's wisdom to his earlier studies in the Arab world, particularly in Damascus. Not only did he learn magic and the Cabala, but also observed that the scholars and wise men freely shared their knowledge – unlike snobbish and buttoned-up Europe. It was in Damascus that he conceived the idea of establishing a fraternity of scholars in emulation of the eastern style of learning when he returned home.

Rosenkreutz, however, was rebuffed when he tried to introduce his idea for a brotherhood of 'magicians, Cabalists, physicians and philosophers' into Europe. So, after a few years back in his native Germany, he decided to form a secret fraternity, beginning with just three followers. The order grew swiftly, devoting itself mainly to healing the sick. Christian Rosenkreutz died at the age of one hundred and six – in 1484 or 1485 – and his burial place was considered lost until a long-hidden tomb was discovered in the House of the Holy Spirit, which the order had built as its headquarters. The discovery, which was – judging from the texts' internal chronology, in 1604 – a vault lit by an 'inner sun' with walls covered in geometric shapes, and which contained all kinds of wonderful instruments and devices, and the founder's body beneath an altar was the sign that the 'general reformation of the world' that Rosenkreutz anticipated was finally at hand.

The brotherhood declared itself to be Christian, but of a reformed kind, and to follow an alchemical philosophy whose focus was on transmuting base souls into divine gold. They firmly rejected the notion that their practice was 'ungodly and accursed gold making'.[5] The *Confessio* declared 'the Pope of Rome Antichrist' in anticipation that the cooperation of the learned would overthrow His Holiness, and by implication the entire Catholic Church. The coming of the 'light of truth' had been heralded by new stars appearing in the constellations of Serpentarius and Cygnus in 1604, which links to the discovery of Christian Rosenkreutz's tomb in the *Fama*. (Kepler also thought that these new stars presaged religious and political changes.)

In 1614 the *Fama* and *Confessio* caused great excitement – and unsurprisingly great hostility from those opposed to such innovations, most obviously Catholics. Tobias Churton calls them 'one of the most virulent intellectual hurricanes ever to hit Europe',[6] while referring to the Rosicrucian furore as Europe's 'first multi-national conspiracy story'.[7] The manifestos announced the existence of a secret, elite brotherhood, which was privy to advanced knowledge, and invited applications for membership – but gave no clue about how to do so, implying that only those capable of working it out were worthy of joining. As a result, interested men of learning started writing their own tracts and open letters to the Rosicrucians, appealing for admission. On the other side, pamphleteers denounced the fraternity as subversive and dangerous, no doubt looking over their shoulders as they did so.

As one of the most effective publicity campaigns in history, the manifestos have been a source of perplexity ever since. Was there really a secret society behind them? Or was the whole point to make people *believe* that such a thing existed? Was it all some kind of hoax? And what was the meaning of the rose and cross symbolism, which has exercised esoteric imaginations ever since? Many suggestions have been made: Martin Luther's emblem was a cross within a rose – and is reproduced in Andreae's *Chemical Wedding*. Yates suggested it could be a combination of two alchemical terms, *ros* (dew) and *crux* (cross).[8] And yet the answer might be much simpler: Andreae's coat of arms was a St Andrew's cross surrounded by four roses.[9] Or perhaps the answer lies in a conflation of all three, since Andreae was a Lutheran and an alchemical influence strongly pervades the manifestos. And while subtlety might be the key to understanding the texts, many commentators over the years have erred on the side of one of the two extremes and have taken everything in the manifestos literally or dismissed them completely as a hoax or fantasy.

Andreae himself often used the term *ludibrium* in relation to the manifestos and the Rosicrucians in general. He also applied this term to his own *Chemical Wedding*. *Ludibrium* basically means a jest, game or a play, which given Andreae's moonlighting activities as a playwright, and his love of the theatre – he particularly admired English drama – probably best describes his intentions. While not literally true, the manifestos were, in Churton's words, 'a dramatic joke with serious intent'.[10] This description calls to mind other similar manifestations, including the relentless social sarcasm of Charles Dickens' comic scenes, the steely undercurrent of today's political satire or, as we claim elsewhere, the subtext behind Leonardo's 'Holy Shroud' of Turin, which we also describe as a *commedia*, or serious joke.

Disappointingly, the story of the great Christian Rosenkreutz in his strangely lit entombment and the origins of his Fraternity are certainly not factually true. After examining Andreae's later voluminous writings, Tobias

Churton proposes that the manifestos are an allegorical account of the transmission of the philosophy that Rosicrucianism continues. Originating in the Middle East, it was preserved in the Arab world before entering Europe via Spain (the *Fama* describes 'C.R.' returning from Arab lands through Spain). But as Andreae decries in other writings, after a promising start that tradition came to a shuddering halt when the brotherhood had to go underground. Now the time was right for it to re-emerge, heralding the coming of a new world fit for heroes.

In the same way that the writers of utopian works, which Rosicrucian were very much in vogue at the time (for example Campanella's *City of the Sun*), hoped to inspire people to attempt to achieve their perfect society, the Rosicrucian manifestos aimed to provoke readers into banding together to create a learned philosophical brother-hood based on the principles they described. Inviting membership was one method to achieve this. By bringing fellow travellers into the open, they could then begin to build their own utopia, completing a self-fulfilling prophecy.

But was there a secret society behind the tracts? Although this question is harder to answer, clearly the publications were part of a campaign organized by a group of like-minded individuals, who we can legitimately call Rosicrucians, if only for want of a better term. As we will see, there is a suggestion that this group called itself 'Antilia'.

However, in answering the above question, let us also not forget about one group in particular. Experienced in operating underground and passionately dedicated to creating a brave new world from its heartland in Lutheran Germany, Bruno's Giordanisti, formed a quarter of a century before, certainly presents itself as a potential candidate for the secret society behind the manifestos. As we will see, there were specific connections between Andreae's circle and the Italian radical Hermeticists connected with Bruno and Campanella, and the Giordanisti would be a natural conduit between the two.

HERMETICISM REPACKAGED

The underlying esoteric philosophy contained in the manifestos was the Renaissance occult philosophy, which as we have seen had Hermeticism at its core. It also highlighted another tradition that had yet to feature prominently in the Hermetic revival: alchemy. A word derived from 'Al Khem', the ancient Egyptian word for their country, 'alchemy' is also the root of the modern word 'chemistry'. Despite being derived from Hermetic principles – essentially their application in the field of chemistry – alchemy had yet to become a major part of occult philosophy, coming into Rosicrucianism through the works of the early sixteenth century physician and esotericist Paracelsus.[11] This is especially fitting given that the Rose Cross' main concern was always healing.

Another Hermetic giant whose philosophy heavily influenced the manifestos was John Dee. His masterwork, *The Hieroglyphic Monad (Monas hieroglyphica*, 1564), presented a new symbol, derived from astrological and other magical glyphs, which he believed embodied the secrets of the cosmos. The significance of Dee's arcane treatise can be deduced from the fact that it was the basis of the Latin tract *A Brief Consideration of a More Secret Philosophy* (*Secretioris philosophiae consideratio brevis*) that prefaced the Rosicrucian *Confessio*. Attributed to Philip à Gabella, who was almost certainly fictitious – his surname probably a reference to the Cabala – it presents explanations, complete with handy diagrams, which shed some light on Dee's distinctly abstruse work. The clear suggestion is that the 'more secret philosophy' behind that penned by the Rosicrucians is Dee's, whose importance to the movement is underscored by the fact that Andreae's *Chemical Wedding* is decorated with his *monas hieroglyphica* symbol.

The legacy of the great English Hermeticist was obviously hugely important to the shadowy occultists behind the Rosicrucian manifestos. This is perhaps not only true in the world of magic, for Dee was a friend of Elizabeth I, besides being her astrologer, spymaster (whose codename was 007) and a major figure behind the explosive expansion of the emerging English Empire. His was a very useful name.

Andreae was a deeply committed Christian – the motto ascribed to the Fraternity of the Rose Cross, and used elsewhere in Andreae's writing, is *Jesus mihil omnia*, 'Jesus above all'. However, in Tobias Churton's words, 'There are clearly many elements of Andreae's thought – not counting his early and fecund immersion in the world of alchemy – which are clearly of Hermetic provenance.'[12] In one of his later works, Andreae praised Pico della Mirandola for being one of the pioneers of the philosophy and spirit that he wanted to see more of, besides lamenting its decline in his own day.

The Hermetic basis of Rosicrucianism can be seen in the works of two major devotees, one English and the other German, who both recognized Rosicrucianism as a development of Hermeticism.

The English physician Robert Fludd (1574–1637) was a major intellect of the period, and like any good Renaissance man was passionately devoted to the pursuit of all knowledge. His work was heavily influenced by – really, a continuation of – that of Pico, Ficino and Agrippa, and he quotes constantly from the *Corpus Hermeticum* and *Asclepius*. There are resonances with Bruno's works that indicate Fludd was familiar with them, although he never mentions the Hermetic martyr directly.

It would be surprising if Fludd had not studied Bruno, since he was a great exponent of the magical art of memory for which Bruno was most famous. In Fludd's version, the basic 'memory buildings', the interior of which the practitioner holds in his or her imagination, mentally placing talismanic images at specific points within them, are conceptualized as theatres. And, it seems, the theatre on which Fludd based his system was none other than Shakespeare's legendary Globe, highlighting the theatrical and dramatic undercurrents that run throughout this story.[13]

Fludd attempted to attract the interest of the Rosicrucians by publishing, in 1616 and 1617, two books on the subject defending them from attack. In both he elucidates his belief that the works of 'Mercurius Trismegistus' are the supreme source of the tradition of ancient wisdom of which he himself and the Rosicrucians were a proud part. He was also a devout Anglican, again showing that Christian piety was considered utterly compatible with the arcane.

Later, in 1633, Fludd was to write that the name of the Brothers of the Rose Cross is 'so odious to contemporaries that it is already buried away from the memory of man'.[14] While some see this statement as repudiation in all but name, the reality is very different. Fludd was actually explaining why the brotherhood changed its name to 'the Wise'. As we will see, by the time Fludd wrote this, the Rose Cross had endured attacks that had given it a dark reputation.

Michael Maier (1568–1622) was a very similar figure to Fludd. A respectable physician and committed Lutheran, he was also a distinguished alchemist. For a time he was also doctor and counsellor to the great esoteric patron Emperor Rudolph II, to whom he dedicated a study of Hermes Trismegistus. From 1611 Maier also spent five years at James I's court in London. Long after his death, his work came to influence the genius that was Isaac Newton. But once again, as Maier was a likely candidate for the Giordanisti, we find the shadow of Bruno towering in the background.

Both Fludd and Maier were dyed-in-the-wool Hermeticists, basing their work firmly on the Hermetic philosophy. This is particularly significant given that they seem to have dismissed Isaac Casaubon's damning historical critique, despite undoubtedly being aware of it. Both moved in the same English intellectual circles as Casaubon, and Maier was even at James I's court when he published his book at the King's instigation.

When we look more closely at the traditions behind the manifestos, and their direct connections with the Hermetic reform movement, it becomes very obvious that Rosicrucianism was a repackaging of the not-very-secret agenda of Bruno and Campanella.

The essential message of the manifestos was that a new reformation was needed. And the increasingly chaotic world in which the manifestos emerged certainly showed that change was needed. The Protestant reformation was failing externally through Catholic pressure as well as through internal division. The Counter Reformation that spawned the likes of the Jesuits was causing great havoc and threatening to take Europe back into the Dark Ages. The situation was slipping out of Protestant control.

The Rosicrucians sought a return to primitive, unadorned and non-popish Christianity, blended with unashamed mysticism and shot through with a kind of spiritualism. They advocated a form of shamanism or mediumship, by which practical and magical information was communicated from the spiritual dimension. Overlying all this, however, was the drive towards self-transformation through alchemy of the body and soul. All things would be possible to the initiate, who was radiant with Christ's love and power and would stride forth into transcendence as a human god. This was the ultimate glittering prize and its seekers would do everything in their power to see that they remained in the race to win it.

It is surely beyond coincidence that the Rosicrucians should emerge in the same circles and espouse the same principles as the Giordanisti that Bruno founded in Germany in the late 1580s and early 1590s, little more than a decade before his death. But there were more direct connections between the Rosicrucians and the Italian side of the Hermetic reform movement. The *Fama* was bound with a German translation of a chapter from the Venetian Traiano Boccalini's *News from Parnassus*, which had appeared two years earlier, calling for, in the words of the *Fama*, a 'general reformation of the whole wide world'. We recall that unsurprisingly the Bruno-inspired Boccalini was an enthusiastic member of Galileo's intellectual circle. This pairing of books links the German Rosicrucian current with 'secret, mystical, philosophical and anti-Hapsburg currents of Italian origin'.[15] As if to remove all doubt of this connection, Andreae defends Boccalini in his *Three Books of Christian Mythology* (*Mythologiae Christianae Libri tres*, 1618).

The conclusive link, however, is found in the two German disciples who visited Tommaso Campanella in prison in Naples and got his books published in Frankfurt. Tobias Adami and Wilhelm Wense were Andreae's close friends and members of the Societas Christiana that he founded in or around 1618. This society embodied the same spirit and principles advocated by the manifestos – religious reform based on the Christian principle of 'love thy neighbour', and the use of scientific enquiry to improve the human condition – but in a more overt and less esoteric way. It was to be the first of a network of Christian Unions, which Adami proposed should be called the City of the Sun, explicitly based on Campanella's as-yet unpublished work of the same name (which Adami finally managed to get published in 1623).[16] *City of the Sun* was also notably a strong influence on Andreae's utopian *Christianopolis* (1619).

Which leaves us with the big question, why choose that particular time to introduce Rosicrucianism to the waiting world?

THE ALCHEMICAL WEDDING

In 1612 James I bequeathed his daughter Elizabeth to the mystical Frederick V, the Elector Palatine, hereditary ruler of the German state of the Palatinate of the Rhine and leader of the Protestant Union, a coalition of Protestant German states formed four years earlier for mutual defence against the Catholic powers. This was seen as a great sign in esoteric circles; it revived those hopes that had once centred on Elizabeth I, Bruno's great goddess, the self-created living icon of the bewigged and jewel-encrusted Gloriana. Her successor James I (of England and VI of Scotland) was notoriously suspicious of all forms of occultism. Upon his succession in the first decade of the seventeenth century, he withdrew royal patronage from Dr Dee, causing a serious decline in the old man's fortunes and a sad slide into death. But the union between James' daughter and the Elector unequivocally aligned England with the Protestant Union, which had a direct political appeal to James. But it was viewed among those hostile to the Church of Rome with a fervour bordering on the apocalyptic.

As the geopolitics of seventeenth-century Europe spiralling into the Thirty Years War often seems like a morass of confusion, it is worth revisiting the Hermetic agenda at that time. Bruno and Campanella had worked to head off what seemed set to become a catastrophic confrontation between the forces of Protestantism and Catholicism by attempting to reconcile each sides' claims to primacy. The Catholic Church claimed the authority of the apostolic succession going back to Saint Peter, while the Protestants, although a new movement, claimed to be returning Christianity to Jesus' original vision. Meanwhile, the Hermeticists, by claiming Egypt as an antecedent to Christianity itself, were trying to offer a middle path to both sides – with astonishing naivety, or so it seems with hindsight. On a more realistic, political level, the Hermeticists plotted to gain influence over the most enlightened monarchs from both camps, for example when Bruno wooed Elizabeth I on one side and Henri III on the other. By the time of the betrothal of James' daughter and the Elector Palatine, however, it was crystal clear that the Catholics, now headed by the Spanish Habsburg monarchy, were in no mood to compromise. So while individual Catholic Hermeticists such as Campanella stuck to their agenda, those on the Protestant side had to work and pray for a more robust counter movement to take shape – until there was another opportunity for reconciliation.

The prospects of a new Elizabethan age, and of a united Protestant Europe, were made more likely by the fact that Princess Elizabeth, who was seventeen at the time of her marriage, was very likely to become queen. The heir to the throne, her older brother Henry, Prince of Wales, had died of fever just a few months before, and her younger brother, twelve-year-old Charles, had been in such poor health since infancy that few expected him to reach adulthood. (As things turned out, Charles did succeed his father as Charles I, but was doomed to be beheaded at the hands of Cromwell's Parliament.)

Frederick came to England at the end of 1612 for the wedding and fell for his bride at first sight. The celebrations ran on for months, extravagant even by the standards of royal weddings. The great poets of the day wrote rapturously of the couple, songs were composed and elaborate masques were written and designed by the greatest names. The celebrated metaphysical poet John Donne wrote of Elizabeth:

> Be thou a new star, that to us portends
> Ends of great wonder; and be thou those ends.

Shakespeare's company, the King's Men, performed a series of plays at court during the months leading up to the wedding. His most overtly esoteric work, *The Tempest* (whose magician character, Prospero, was allegedly based on Dee) was performed on the betrothal night, 27 December 1612, with some additional scenes specially written for the occasion.

With a deft PR touch Frederick and Elizabeth of Bohemia were married on Valentine's Day 1613, after which the couple went to live in the romantic Heidelberg Castle in the Palatinate. Frederick had constructed for his love what was regarded as the eighth wonder of the world, the famed *Hortus Palatinus*, an Italian Renaissance-style landscaped garden, laid out in deeply symbolic fashion, complete with mechanically animated statues, imported tropical plants and a celebrated water organ.

The appearance of the Rosicrucian manifestos in the two years immediately following the wedding was intimately bound up with the expectations of esoteric reform that centred on the couple. It was probably no coincidence that the Palatinate bordered on the Duchy of Würtemmberg, home to Johann Valentin Andreae. The works were preparing philosophical circles in Germany and beyond for the new era that they believed this golden couple would usher in – a unified Protestant Europe that would confront the ultra-Catholic nations.

Other events underscore the connection between the marriage and the manifestos. Robert Fludd's works of that period were published in two volumes in 1617 and 1619 in the Palatinate (despite being written in England), as were

magical brotherhood – *sorcerers*, no less – abroad in the city, up to God knows what and only God would know what because they were invisible. Books and pamphlets speedily appeared warning that the Invisibles were part of a devilish plot. The anonymous but presumably delightful *Horrible Pacts made between the Devil and the Pretended Invisible Ones* claimed that the Invisibles were part of a global Satanic conspiracy, that six groups of six members in different areas around the world were plotting mankind's downfall. Another pamphlet specifically named Michael Maier as their leader. The Jesuit François Garasse called them 'a diabolical secret society who should be broken on the wheel or hanged on the gallows'.[17]

If this seems all rather sensational, then no doubt that was the intention. After all, claiming to be invisible Rosicrucians was likely to provoke over-heated imaginings. The PR genius involved in whipping up this type of frenzy suggests that the notices were actually the work of Rosicrucian haters, or more accurately enemies of Rosicrucianism.

Why should anyone want to stir up anti-Rosicrucian paranoia, especially at that particular place and time? As Parisian intellectuals became fervently hooked on the manifestos' furore and the works of their defenders such as Michael Maier, generating a major scare would have acted like a cold shower on potential new devotees. If all the hot air about pacts with the devil gave the impression that to dabble in Rosicrucianism would guarantee an eternity of being prodded by poker-wielding demons, then a similar fate would surely await them whilst they were still alive, care of the Pope's men.

It is unlikely to be a coincidence that in the same year the Hermetic tradition also came under a sustained onslaught in Paris, from Marin Mersenne, a Jesuit-educated monk. In works published from 1623 he attacked everyone from Pico della Mirandola onwards, reserving special hostility for

Robert Fludd, with whom he engaged in a high-profile war of words. He wrote of Bruno that he had 'invented a new way of philosophy in order to secretly fight against the Christian religion'.[18] And tellingly, Mersenne was the first to use Casaubon's re-dating of the Hermetica against its devotees.

'FROM MAGIC TO MECHANISM'

The Invisibles scare and Mersenne's onslaught on Hermeticism also forms the unexpected backdrop to the rise of *the* arch-rationalist of the time, he who also set the tone for the coming age of science. The philosophy of René Descartes (1596–1650) would ultimately lead to the divorce between the magical and scientific components of the Hermetic tradition. But, importantly, his career also demonstrates how this divorce was due as much to the exigencies of the time as it was to a change in the intellectual direction.

Descartes was the Jesuit-educated French philosopher who argued that all physical phenomena could be reduced to, and explained in, purely mechanical terms. 'Cartesianism' represented a 'shift from magic to mechanism'.[19] But he also introduced the idea of a duality between mind and body, the consequence of which we are still coming to terms with.

Although his work is usually portrayed as a reaction against religion, the target of his argument was actually Rosicrucianism. Descartes was certainly not viewed as an enemy of Catholicism at the time. Quite the opposite – his ideas were actively encouraged by at least one leading Catholic theologian because of the ammunition they provided the Church in its onslaught against Rosicrucianism and Hermeticism.

Indeed, in his young days Descartes had been something of a Papist swashbuckler; as a twenty-four-year-old he had fought with the Catholic forces at the Battle of the White Mountain of 1620 that smashed the hopes of the Alchemical

Wedding. He entered Prague with the victorious troops. It was when quartering during the long months of the previous winter that he heard talk of the Rosicrucians and – perhaps oddly for a Pope's man – found himself interested in them. Realizing that the Fraternity's ideals and principles chimed with his own developing ideas, Descartes tried to make contact with it. He failed, but while he was holed up at Ulm in the summer of 1620 he met the mathematician Johann Faulhaber, who had tried to approach the Fraternity as a would-be member and had some useful knowledge to share with him.

Descartes returned to Paris in 1623, and found himself in the middle of the 'Invisibles' scare. This threatened to be somewhat dangerous for him given that it was known that he had been interested in the Rosicrucians while in Germany. As the anti-Rosicrucian hysteria was threatening to turn into a lynch-mob scenario, to save his skin, Descartes made a point of denouncing the Rosicrucian 'calumny'.

As we saw earlier, in the vanguard of the opposition to Rosicrucianism in Paris and beyond was the monk Marin Mersenne (1588–1648), of the exquisitely named Order of Minims. As we have seen, he was the first to use Isaac Casaubon's re-dating of the Hermetica against the likes of Robert Fludd. Eight years Descartes' senior, he had been a fellow student at the Jesuit college at La Flèche in the Loire, and the two men were close friends.

Besides being a theologian, Mersenne was a mathematician and scientist, best remembered today for his work on acoustics and prime numbers, a dubious combination of interests for a devout Catholic at the time, as the Counter Reformation, and particularly the overthrow of Frederick V, had invested these subjects with a heavy taint of occultism. Mersenne was eager to rescue his fields of interest from any suggestion of diabolism. In 1623, the year of the 'Invisibles' scare, Mersenne published *Famous Questions in Genesis*

(*Quaestiones celeberrimae in Genesim*), which despite its title was a fierce attack on the occult philosophy and its advocates, especially Pico della Mirandola, Ficino, Agrippa and, particularly, Robert Fludd. On the other hand, he was a defender of Galileo, and expressed some admiration for the intellect of Campanella, who he met in Paris when the Italian was under Cardinal Richelieu's patronage, although he dismissed his philosophy outright.

To Mersenne, Descartes' concepts were potentially an excellent way of ridding natural philosophy of any suggestion of the esoteric, so he encouraged him to publish and helped promote his work, effectively acting as his agent for his first book, *Meditations on First Philosophy* (*Meditationes de prima philosophia*). Ironically, the full title was *Meditations on First Philosophy, in which is Demonstrated the Existence of God and the Immortality of the Soul* – Descartes wasn't as extreme a rationalist as he is often portrayed today. In fact, given Descartes' religious beliefs and Mersenne's support, the Cartesian revolution was, if anything, a Catholic reaction against Rosicrucianism and Hermeticism.

After Descartes, natural philosophy bifurcated into two camps, each advocating a different way of acquiring knowledge. There was the mechanist philosophy, in which everything could be reduced to and understood in terms of physical properties – the characteristics of bodies and the forces that act on them. On the other side was the Hermetic approach, which saw things more holistically, every imaginable thing being inextricably part of a great living whole. Ultimately, of course, mechanism won the day, although it was by no means an overnight victory.

With the rise of Descartes' influence, the philosophy that had driven the Renaissance was at its lowest ebb, seemingly heading for complete extinction. In half a century Isaac Casaubon had challenged it historically, the Thirty Years War had dashed its hopes politically and now Descartes was undermining it philosophically. But this was not the end of the story. There were those who kept the Hermetic torch alight, even in the heart of Rome itself. And it was yet to see its greatest triumph in the scientific world.

CHAPTER FIVE

SIGNS, SYMBOLS AND SILENCE

One might be forgiven for thinking that as the Age of Enlightenment moved inexorably towards the Age of Science, Hermeticism was, if not actually dead then pretty much moribund. But in fact, for the most part, it just continued in disguise. For obvious reasons of self-preservation after the polarization of the Thirty Years War, most thinkers who were inspired by the Renaissance occult tradition downplayed that fact, while quietly continuing on their path. Others, meanwhile, took little care to be circumspect, and astonishingly, got away with it. These two approaches – covert and overt – were respectively adopted by two of the seventeenth century's most remarkable minds: Gottfried Wilhelm Liebniz and Athanasius Kircher.

TRUE CABALA
Leibniz (1646–1716) vies with his exact contemporary, Isaac Newton, for the title of the century's greatest intellect. His output covered every conceivable field of his day, from linguistics through engineering to biology, his mind leaping chaotically from subject to subject. In his lifetime he published about a dozen works, but most of his thoughts, ideas and discoveries were scattered in a vast number of papers, letters and half-completed books, the majority of which have yet to be published. Yet we do know something

particularly significant about Leibniz: he was heavily rumoured to be at the very least a Rosicrucian sympathizer.

Leibniz's major contributions were in the increasingly important fields of mathematics, logic and metaphysical philosophy. As he devised infinitesimal calculus in the late 1670s, at the same time as Newton, a protracted row between the two men erupted, with Newton accusing Leibniz of stealing his invention. In the end it was Leibniz's notation that became the standard. He also invented the binary system on which our digital world depends and without which, in fact, most of the modern world could not exist.

Like many intellectual giants of the time, Leibniz's career was an odd mix of science, philosophy and diplomacy. While working for Georg Ludwig, Duke of Brunswick, he even got involved in negotiations over the English Act of Settlement of 1701. This Act bestowed the crown on the descendants of the Duke's mother Sophia, establishing the run of over-stuffed and not always totally sane Hanoverian Georges on the British throne. Sophia, to whom Leibniz was mentor and adviser, was the Electress of Hanover and daughter of the Winter King and Queen, Frederick V and Elizabeth Stuart. And so we find a rumoured Rosicrucian working for the family of the alchemical bride and groom – suspiciously neat.

Born in Leipzig, Leibniz's first job after receiving his doctorate in law was as an alchemist in Nuremberg, where he was rumoured to have joined a Rosicrucian society. There is probably some substance to the story, which was accepted, for example, by the French mathematician Louis Couturat, author of a 1901 study of Liebniz.[1] There are potential Rosicrucian connections with Nuremberg: in 1630 Johann Valentin Andreae tried to revive his Societas Christiana in that city, so there may still have been a coterie of fellow travellers there three decades later.

Not only did Leibniz practise as an alchemist, but his later works reveal a deep familiarity with the Rosicrucian manifestos and with Andreae's writings. He proposed the formation of an Order of Charity and drew up its constitution – part of which is lifted directly from the *Fama Fraternitatis*.[2] So at a conservative estimate, Leibniz certainly had Rosicrucian leanings.

Leibniz's first major work, *Dissertation on the Art of Combination* (*Dissertatio de arte combinatoria*), published in 1666 when he was just twenty, is about the art of memory – although the non-occult version, simply as an aid for remembering. In the introduction, he acknowledges his debt to previous practitioners such as Bruno, and goes so far as to lift the term *combinatoria* from him.[3]

But did Leibniz, as many historians assume, completely abandon these interests when he realized mathematics and logic were the way forward? Certainly Leibniz's career did seem to be set to embrace all things mechanistic. He devoted himself to absorbing the latest thinking – including certain of Descartes' then-unpublished writings – during a four-year sojourn in Paris on a diplomatic mission for the Elector of Mainz. During that time, in 1673, he took a trip to London, where he wowed the Royal Society with his innovative calculating machine and was duly made a Fellow.

But later Leibniz realized that the mechanistic approach was limited, writing to a correspondent two years before his death:

But when I looked for the ultimate reasons for mechanism, and even for the laws of motion, I was greatly surprised to see that they could not be found in mathematics but that I should have to return to metaphysics.[4]

Any search for the source of Leibniz's metaphysical inspiration begins with his devotion to Marsilio Ficino's 'perennial philosophy'[5] – Hermeticism. Bruno's influence, too, filters directly through to Leibniz, possibly through the conduit of the Giordanisti.

Leibniz's search for a metaphysical explanation for 'ultimate reasons' led him to formulate his theory of monads, which, to put it politely, is a somewhat abstruse idea. His monads are a kind of metaphysical or spiritual equivalent of atoms, the indivisible building blocks from which everything in creation is comprised and which are attached to physical atoms. Monads all originated at the beginning of the universe and, since they can neither be created nor destroyed, all change consists merely of their transformation.

Monad is the Greek word for unity, and since the time of the Greek philosophers it has been used to describe basic units and first causes in many different philosophies – it is an important concept in Neoplatonism, for example. Leibniz's concept of monad, however, was directly influenced by Bruno.[6] As Frances Yates pointed out:

> Though Leibniz as a philosopher of the seventeenth century has moved into another new atmosphere and a new world, the Leibnizian monadology bears upon it the obvious marks of the Hermetic tradition.[7]

In the interests of self-preservation, Leibniz himself was reluctant to acknowledge the influence of the Hermetic tradition. On the one hand, in the volatile new climate after the Thirty Years War Hermeticism was tainted with the whiff of heresy and diabolism, almost entirely because of Bruno. On the other hand – and partly as a consequence of being tainted – the reputation of Hermes' system in scientific and intellectual circles had suffered, and it was beginning to look old-fashioned and misguided.

But even if Leibniz was wary of shouting it from the rooftops, his works quite clearly owe a major debt to the Renaissance occult philosophy. Even Leibniz's system of calculus evolved from this tradition. It developed from his quest to reduce everything, not just scientific principles and laws but also religious and ethical questions, to a common symbolic language: a universal calculus. Building on the art of memory, both the classical and 'occult' versions, in order to establish a language of symbols or *characteristica universalis*, Leibniz envisaged a set of images to which all the fundamentals of knowledge could be reduced. This naturally necessitated the cataloguing and codification of all that was known, a growing eighteenth-century preoccupation. By manipulating and setting the symbols in different relationships, he believed that new discoveries could be made.

He specifically likened such a system to Egyptian hieroglyphs, which along with Bruno, he believed were used in a similar way. Leibniz also considered, but eventually rejected, Dee's innovative *monas hieroglyphica* symbol. The Cabala, too, was an influence, since it is based on the idea that certain principles are present in all things. Leibniz even described his *characteristica universalis* as 'true Cabala'[8] – hardly the words of a modern-style rationalist.

Eventually Leibniz came to realize that the best tools for the job were mathematical symbols. This realization then led to the development of his version of infinitesimal calculus, which he intended to be a first step towards the universal calculus. Although Liebniz developed his concepts in a mathematical and mechanical direction, in focusing on a universal calculus he was closely following Bruno, who had extended the esoteric art of memory to include complex techniques for combining the images held in the mind in different ways.

In addition to his formulation of the binary system, in this mode of thinking Leibniz was anticipating modern computer modelling, which is based on the idea that any system can be defined in mathematical terms, reduced to values, variables and relationships that can be manipulated in the computer to predict how the system will behave under varying conditions. Leibniz laid the foundation for contemporary information theory, and also saw the potential for creating machines to do the hard work of combining his *characteristica universalis*. Not only did he invent mechanical calculating machines that could do basic arithmetical operations, but he also tried to design one for more complex algebraic calculations. He even conceived a device that used binary mathematics.

Is it a stretch to say that mathematical equations, the modern scientific use of formulae and even some of the basics of computer science comes from an occult idea? Clearly Leibniz himself saw his work that way, even defensively describing the *characteristica universalis* as 'innocent magia'.[9] There is no denying Leibniz's unique contribution to mathematics and computer science – but significantly it may also be fair to say that these were largely inspired by the Hermetic tradition.

EGYPT'S LAST STAND

In the midst of all the Hermetic reversals in fortune there seems to have been a last and perhaps desperate attempt to carry the tradition into the very heart of Rome in a way that would have made Giordano Bruno very proud.

As we saw earlier, in the 1580s Pope Sixtus V had presided over the raising of an ancient Egyptian obelisk in Saint Peter's Square to mark the final trouncing of paganism. But there was another episode of obelisk-raising in the 1650s and 1660s that was motivated by the exact opposite. Inspiration for the second wave can be traced back almost entirely to one man, another acknowledged genius of the age, one of those paradoxical figures who according to the usual simplistic view of the period should not really have existed: the extraordinary Hermetic Jesuit Athanasius Kircher.

Kircher was a polymath and gifted mathematician – he has been called the 'last Renaissance man' and 'the last man who knew everything' – and is regarded by many as the founder of Egyptology. He was born in either 1601 or 1602 (he didn't know which, although happily he knew his birthday) in Hesse-Cassel in Germany. After being educated at the Jesuit College in his hometown of Fulda, he entered the Society of Jesus in 1618.

There is no way that Kircher could not have been aware of the furore over the Rosicrucian manifestos. Not only were they were published in Hesse-Cassel and widely debated during the 1610s and 1620s, but also the Jesuits spearheaded the opposition to them. And all the key Rosicrucian elements turn up in Kircher's works – everything but the name, in fact.

In 1631, during the Thirty Years War, Kircher was forced to flee, swimming across the Rhine to escape Protestant forces. He made his way to Avignon where he taught mathematics in the Jesuit College, before becoming a professor of mathematics at the Society's most prestigious establishment, the Collegio Romano in Rome. By that time he was widely thought of as a brilliant mathematician and polymath, and had gained the confidence of the Pope. As befitted the 'last Renaissance man', Kircher studied medicine, besides being a great inventor and a musician. He experimented with the magic lantern and the projection of images. He was a geologist and fossil-collector whose intellectual curiosity was so great he ventured into the crater of Mount Vesuvius when it was threatening to erupt. Perhaps as a result of an association of ideas, he also designed firework displays. By any standards, Kircher's career was extraordinary. So much so, in fact, that in 2002 a number of distinguished scholars convened at the New York Institute of the Humanities to debate 'Was Athanasius Kircher the coolest guy ever, or what?' They concluded that he was, no 'whats' about it.[10]

His work with microscopes led him to argue that little 'worms' propagate plague, the earliest statement of the germ theory of disease based on microscopic observation. He also calculated that the height needed for the Tower of Babel to reach the Moon would knock the Earth off its path through the skies, which was particularly interesting as he shouldn't have acknowledged that the Earth had an orbit in the first place! He argued that animals would have had to adapt to life after the Flood, one of the first recognitions of evolution. But like Leonardo before him, there was an element of the joker about Kircher. He launched little hot air balloons with 'Flee the wrath of God' written underneath, and dressed cats up as cherubs. He also designed – but mercifully probably never built – a katzenklavier, a musical instrument that produced a range of sounds when a semi-circle of cats had pins stuck in their tails. It was clearly not a good idea to be a cat around Kircher.

But most of all, he was passionate about ancient Egypt. To him, deciphering hieroglyphs would reveal the language that God gave to Adam, bestowing all the secrets of the universe. Indeed, thanks to a book that Kircher encountered in the Jesuit college library during his training, his main obsession was hieroglyphs – which nobody could read then (and wouldn't until the discovery of the Rosetta Stone in 1799). Kircher became convinced he had made the longed-for breakthrough required to crack the code, although we now know that this was wishful thinking. However, his passion is one of the reasons he was so enthusiastic about getting involved in the re-erection of obelisks, as he lusted after the opportunity to study their inscriptions. While he was a professor in Rome, Kircher even dispatched a student to Egypt to measure the Great Pyramid, inside and out, and to copy hieroglyphs from two standing obelisks in Alexandria and Heliopolis – probably not the quickest or easiest assignment the young man had ever been given.

Like most scholars in those days, Kircher was convinced that the hieroglyphs inscribed on temples, statues and obelisks embodied the wisdom and science of ancient Egypt. Surely it would only be a matter of time before a genius such as himself would claim to be the first to understand it all? In Avignon he benefited from a friendship with the astronomer and antiquarian Nicolas Claude Fabri de Peiresc, who had not only travelled to Egypt, but had also brought back various relics. As a fellow astronomer, Fabri de Peiresc was one of Galileo's correspondents who leapt to his defence. Less understandably, he also publicly defended Tomasso Campanella.

Kircher's interest in the mysteries of Egypt naturally brought him into contact with Hermeticism, for which he made no attempt to hide his enthusiasm. But he completely ignored Casaubon's dismissal of the antiquity of the Hermetica, arguing that the texts represented the authentic philosophy, cosmology and religion of his beloved ancient Egypt. In fact, he not only tried to decipher and translate hieroglyphs but also attempted to relate them to the teachings of the Hermetica. He regarded Hermes as the inventor of hieroglyphs, and the inscriptions on the obelisks as the keys to unveiling his wisdom. He even called the Egyptian looped cross, the *ankh*, the '*crux Hermetica*,' or Hermetic cross.

Kircher was also an astronomer, and while he privately accepted Copernicanism, he was careful to state in public that he denied 'both the idea of the mobility of the earth, and of the inhabitants of the other heavenly globes'.[11] The last part of this refutation suggests that it was Bruno, rather than Galileo, who he had in mind. In fact, Kircher's work often displays such close parallels with Bruno's that he *must* have read his works. It is hard to find anyone more in tune with Bruno's thinking: Kircher, too, believed his religion of Catholicism was heir to the Egyptian tradition, and he took Bruno's cosmology as the basis for his own. For obvious reasons, however, it would not have been a great idea to make this too obvious.

Kircher wrote voluminously, his masterwork being the four-volume *Oedipus Aegypticus*, published between 1652 and 1654, which contains a synthesis of all mystical and esoteric traditions, with Egypt squarely positioned as their foundation. And naturally, he acknowledged the significance of the name of the sacred city of the Egyptians, Heliopolis, City of the Sun.

Kircher greatly admired the ancient Egyptian civilization, upholding it as the ideal model for both politics and religion. This understanding is very close to Bruno's vision of Egypt – dangerously so, one might have thought, for a Jesuit working at the very epicentre of Catholicism in Rome. After all, it does seem a perilously short step from believing that Ancient Egypt is the perfect model to advocating the reform of religion and state to match.

Kircher's other beliefs included the idea that Moses had been schooled in the religion of Egypt, which he had then passed on to the Israelites, who subsequently corrupted it. Again, this is dangerously close to Bruno's thinking. The suggestion here is that, given it was believed that Jesus had been sent to put the Jews back on the right track and to open their religion to the rest of the world, then he was actually restoring the Jewish religion to its *Egyptian* roots. Kircher never made this line of thinking overt – after all, he of all people was no fool.

Not only is *Oedipus Aegypticus* liberally studded with quotes from the Hermetica, but Kircher takes both Hermes Trismegistus' authorship of those books and his antiquity for granted, believing him to be a contemporary of Abraham. He includes a hymn of Hermes from *Pimander*. To this he added, in the words of Peter Tompkins, 'a hieroglyph enjoining silence and the secrecy concerning these sublime doctrines – the colophon employed by the Brothers of the Rosy Cross!'[12] More overtly (and bizarrely), Kircher placed great importance on John Dee's *Monas hieroglyphica*, from which he frequently quotes, linking the symbol to the Egyptian *ankh*.

We can see that Kircher shared exactly the same ideals and influences as the authors of the Rosicrucian manifestos, which seems decidedly odd given that the Rosicrucian movement was a Protestant expression of the Hermetic reform agenda. However, as the Hermeticists were working behind both Protestant and Catholic lines for a common cause, even a Catholic Hermeticist like Kircher would share a similar mindset with the Rosicrucians. Kircher's German background even suggests the possibility of a connection with the Giordanisti.

Maybe Kircher was trying to change Catholicism from within, reviving the old dream of Bruno and Campanella. This is by no means just idle speculation, as he managed to interest two popes in Egyptian ideas, and his work with the ancient obelisks points to more than a mere academic interest in those monuments. In this Kircher collaborated with his great friend, the artist, sculptor and architect Gianlorenzo Bernini (1598–1680), who is most famous for designing St Peter's Square with its magnificent colonnades. (Or rather, he is probably most famous now for featuring so prominently in Dan Brown's *Angels and Demons*.)

Unsurprisingly, Kircher and Bernini's joint projects incorporated a wealth of Egyptian symbolism and motifs, which Bernini incorporated into his other works. George Lechner, an expert on magical and astrological symbolism in Renaissance art – a real-life version of Dan Brown's character Robert Langdon – acknowledges that Bernini's use of Egyptian motifs probably derives from the Hermetica.[13] Kircher and Bernini first worked together on a project, later abandoned, to re-erect a 40 foot (12 m) obelisk that had been discovered in a vineyard. Cardinal Francesco Barberini, the nephew of the then-Pope, Urban VIII, sought to set it up in his palace gardens.

Kircher and Bernini conceived that the base of the monument should feature a life-size sculpted elephant, which would bear the upright obelisk on its back. But what did the elephant signify? Was it simply an error – did Kircher and Bernini perhaps believe elephants came from Egypt? The answer reveals something important about the men's otherwise concealed attitude to their religion.

In the twentieth century the Italian painter Domenico Gnoli, among others, identified the inspiration as an image in the allegorical and highly symbolic book *Poliphilo's Strife of Love in a Dream* (*Hypnerotomachia Poliphili*), published in Venice in 1499, an identification accepted by the American art historian William S. Heckscher.[14] The romance is anonymous, although the first word in each chapter spells out a sentence containing the name 'Frater Franciscus Columna', apparently the name of a Dominican monk in Venice. Despite this clue, other authors have been suggested, including Lorenzo de' Medici and Leon Battista Alberti, the polymath mentor of Leonardo da Vinci.

The tale describes a dream within a dream in which Poliphilo ('Lover of many things' or 'Lover of Polia') searches for his *amorata* Polia, who has rejected him. Inevitably, throughout his adventures he encounters many strange creatures along the way, all illustrated by superb woodcuts. *Hypnerotomachia Poliphili* has exerted a hold over the esoteric imagination to this day, as it seems to convey a profound, if elusive, something in symbolic form. Decoding its hidden message provides the central plot of the 2004 bestseller *The Rule of Four* by Ian Caldwell and Dustin Thomason, and it is mentioned in Roman Polanski's powerful and unsettling movie *The Ninth Gate* (and in the novel on which the movie is based, *The Dumas Club* by Arturo Pérez-Reverte).

In the story of Poliphilo the obelisk on the back of a stone elephant is not only described but also illustrated by one of the woodcuts. Before Gnoli identified it as Bernini and Kircher's joint inspiration, *Hypnerotomachia Poliphili* was already acknowledged by researchers of the esoteric as a major influence elsewhere. An Italian writer on Rosicrucianism, Alberto C. Ambesi, considered that it 'marks the true birth of the Rosy Cross, but in code'.[15] This was not to suggest that either the fraternity or the group that produced the manifestos existed in Venice in 1499, but that the currents of esoteric thought that came together in *Hypnerotomachia Poliphili* later influenced the Rosicrucians.

Although the obelisk-on-an-elephant project was aborted, the idea resurfaced in Kircher and Bernini's last collaboration.

Shortly after Pope Innocent X's election in 1644, Kircher proposed that another recently unearthed obelisk, broken into four pieces, should be re-erected in his honour. The first-century emperor Domitian originally commissioned the 55 foot (16.5 m) obelisk (its height was nearly doubled by Bernini's elaborate fountain base) for Rome's Temple of Serapis. Innocent agreed to Kircher's proposal and put him in charge of the project, again working with Bernini. Together they reassembled the obelisk, Kircher designing the missing pieces – complete with inscriptions – and it became the centrepiece of the elaborate statue-covered Fountain of the Four Rivers in the Piazza Navona, which took until 1651 to complete. The obelisk was topped not with a cross, as one would expect, but a dove, which wasn't a reference to the Holy Spirit or dove of peace but to the emblem of Innocent's family, the Pamphili.

Kircher's own account of the raising of the monument, *Obeliscus Pamphilius,* begins with the mysteries of Egypt, and in particular the secrets of the hieroglyphs, but is again heavily studded with Hermeticism and even includes a lengthy discussion of John Dee's *Monas hieroglyphica.* Incongruous to say the least for a book by a Jesuit commissioned by the Pope himself!

Obeliscus Pamphilius can be said to conceal almost as much as it reveals, and there is a strong suggestion running throughout that Kircher is still hiding something. The frontispiece has occupied esotericists and art historians for generations. In front of a fallen obelisk the winged Mercury (i.e. Hermes) hovers holding a scroll inscribed with hiero-glyphs in front of a woman who represents Kircher's muse. She rests one foot on a cubic block of stone, on which Egyptian tools that are clearly the equivalent of the square and compass of the classic Masonic symbol are inscribed. This is most odd – historically and geographically Masonic symbols should not have been in Rome at that time. The frontispiece also features a cherub holding one finger to his lips. Tod Marder, professor at the State University of New Jersey and a fellow of the American Academy in Rome, a specialist in the works of Bernini, writes:

> Cabalistic in the extreme, Kircher claimed to be purposefully obscuring the real meaning of the obelisk, lest he deprive some other erudite soul of the enlightenment that comes from personal decipher-ment. Kircher wrote a book about the Pamphili Obelisk, as it was called. On the title page appears a little cherub with his forefinger raised to his lips to

signal silence – if you know the secrets herein, it seems to say, keep them to yourself.[16]

The symbolism of the frontispiece is obvious: through the inspiration of Hermes, Kircher is seeking to restore the great Egyptian secrets.

Kircher's charmed life continued when Innocent X died in 1655 and Fabio Chigi was elected as Alexander VII. Alexander was responsible for commissioning Bernini to remodel St Peter's square, with the Caligula obelisk as its centrepiece. Peter Tompkins describes the new Pope as:

> an Hermetic scholar who took a personal interest in Kircher's hieroglyphical studies, contributing generously to the publication of Kircher's many more works, and so, indirectly, to keeping alive the wisdom of Ficino, Pico, and their Thrice Great Master.[17]

The year of Alexander's election was also remarkable for a great discovery. During the digging of a new well, a smallish, 18 foot (5.5 m), pink granite obelisk in good condition was found in the garden of Santa Maria sopra Minerva, a remnant of the original temple to Isis. Having solicited the commission from Alexander, Kircher and Bernini had this set on the back of a stone elephant in the piazza in front of the basilica. With this accomplished, the statue was instantly recognisable as the outward and visible form of the woodcut from the *Hypnerotomachia Poliphili*. It appeared that Kircher and Bernini had finally manifested its extraordinary symbolism in hard stone.

The obelisk is topped with a small and discreet cross, as opposed to the unmissable ironmongery that Sixtus V set on the Caligula obelisk. This one, however, is devoid of smug trumpeting about the victory of Christianity. Instead it provides a perfect reflection of Hermeticism such as

Bruno's, proclaiming as it does that Christianity is built upon and supported by the ancient Hermetic religion of Egypt.

The obelisk's inscription is remarkable because it features a Pope honouring Isis, in what is perhaps an echo of the decorations that could be found in the Appartamento Borgia two centuries earlier: 'Alexander VII erected this obelisk once dedicated to the Egyptians' Pallas [Isis], to the divine wisdom and to the deipara mother.'[18] 'Deipara' means 'mother of God' (so 'deipara mother' is either tautology or heavy emphasis), and is an official Catholic title for the Virgin Mary. But why not make a more explicit reference to Jesus' mother, if she is being honoured here? Clearly because it is referring to Isis, not the Virgin Mary.

To Hermeticists, Santa Maria sopra Minerva was nothing less than a sacred site. Although outwardly a Dominican basilica, it was also the spot where Bruno was taken before his execution and where Galileo abjured his heliocentric beliefs. So here we have an obelisk made in honour of Isis, raised again as part of a Rosicrucian monument by an adherent of Hermeticism, outside the place where Bruno had been condemned and Galileo forced to recant. This was not your average Catholic statue.

But there is still more to this elephantine sculpture, which recalls to us at least the quite jaw-dropping symbolism Leonardo built into his *Virgin of the Rocks*, which we discuss elsewhere.[19] As Peter Tompkins notes gleefully:

> . . . the satirist Segardi, taking the symbolism one step further, used the fact that the elephant's rear end is turned towards the monastery of the Dominicans to compose the epigram, '*Vertit terga elephas vertague proboscide clamat Kyriaci Frates Heid Vos Habeo*' or, in short, 'Dominicans, you may kiss my arse!'[20]

Few are afforded the opportunity to make such extrava-gantly heretical gestures, and indeed this was a last hurrah for Kircher and Bernini. When Pope Alexander died in 1667, Kircher lost papal favour and patronage and resigned from the Jesuit college to concentrate on his intellectual pursuits. In particular he wanted to create a museum preserving artefacts (such as a lizard encased in amber) which he had collected and which Jesuits sent from around the world. With what seemed like destined precision, he and Bernini died on the same day, 28 November 1680.

With the huge confidence (many would say over-whelming arrogance) of a gifted polymath, Kircher, the self-declared Hermeticist and probably a closet Rosicrucian, worked right in the heart of Catholicism, hidden in plain sight. Had he attempted to carry out Bruno's apparently impossible idea of celebrating the compatibility between Hermeticism and Christianity? Surely strangest of all is his success in managing to operate within the rabidly anti-Hermetic and Rosicrucian-hating Jesuit order.

Of course many readers will have noticed that this late flowering of Hermeticism within the Vatican is echoed in the plot of Dan Brown's second novel *Angels and Demons* (2000), as well as in its action-packed movie adaptation. In fact it was returning to the subject of Bernini and Kircher for our roles as contributors to the truth-behind-the-fiction TV documentary tie-in to the movie that led us to unravel many of the above connections.

The fictional basis of Brown's thriller is the supposed existence of a secret society of scientists and freethinkers called the Illuminati, created in the face of persecution by the Church and which boasted Galileo as a prominent member. Because of persecution, particularly Galileo's, the Illuminati became rabidly anti-Catholic, eventually seeking to bring down the Church, which they had infiltrated. One of their secret grand masters was Bernini, who had encoded certain of his Roman works with directions to guide initiates to the Society's hidden base. The hero, Robert Langdon, has to follow the 'Path of Illumination' against the rapidly ticking clock in order to avert an enormous cataclysm and save the day.

As with *The Da Vinci Code*, Brown's grasp of history in *Angels and Demons* has been roundly criticized for its inaccuracies and anachronisms. On the surface, it does seem that liberties have been taken with the facts. Although there was a real secret society called the Order of the Illuminati, whose aims were roughly similar to the organization in *Angels and Demons* – the advancement of free-thinking and the overthrow of the Catholic Church – it wasn't formed until 1776 and was only active in the state of Bavaria. So on geographical and chronological grounds it was impossible for Galileo and Bernini to have been part of it. Critics also focused in particular on the choice of Bernini as the secret Illuminati master, on the grounds that he was a dedicated Catholic who worked for most of his life under the patronage of popes.

The essentials of Dan Brown's story do fit with our own reconstruction, however. If you replace the Illuminati with the Giordanisti then the plot falls into place very neatly, as the latter secretly encouraged scientific thinking and aimed at either the radical reform or overthrow of the Catholic Church. And the Giordanisti was connected with Galileo and, through Kircher, to Bernini. Certain works of Bernini's that Brown used as a framework for Robert Langdon's apocalyptic trip to Rome, such as the Fountain of the Four Rivers in the Piazza Navona, one of the landmarks of the Path of Illumination, are also significant in our own version of the story. So if we substitute the 'Giordanisti' or 'Rosicrucian' for the Illuminati in Brown's novel, we see that, perhaps surprisingly, there *is* a solid historical basis for *Angels and Demons*. (And perhaps it is significant that

Bruno's *On the Heroic Frenzies* culminates in a scene in which nine blind men receive not just sight but insight, becoming the nine 'Illuminati'.) It seems that Dan Brown tapped into a rich vein of synchronicity and serendipity that sometimes, somehow makes life-follow-art-follow-art.

But what of the objection that Bernini was too Catholic to be involved in such shenanigans in the first place? Was he just an innocent fall guy for Kircher's secret Hermetic agenda, as some have suggested? Neither of these objections stand up. As we have seen, even certain popes were devotees of Hermes, and strong Christian beliefs – be they Catholic or Protestant – presented no obstacle to developing an enthusiasm for the works of Thrice Great Hermes. Unless Bernini lived in a bubble and never actually read Kircher's books, he must have understood that the symbolism of their joint works was unequivocally Hermetic.

More importantly, Kircher showed that Bruno's intellectual legacy was not only still alive but also still shaping the development of science. Ingrid D. Rowland, art historian and Fellow of the American Academy in Rome, writes:

Kircher's cosmology and its attendant concept of a universal *panspermia* . . . show that however dramatically the eight-year trial and gruesome public execution of Giordano Bruno had been designed to prove that the heretic philosopher was a lone and terrible fanatic, the performance had failed. Bruno's books had been read by Kepler, Galileo and Athanasius Kircher, and they were enough to change the course of natural philosophy. For both Bruno and Kircher argued with passionate eloquence that nothing but an infinite universe did justice to an omnipotent God, and once the idea of that vastness immeasurable had been conceived, it really did burst the crystalline spheres of Aristotelian physics.[21]

But Hermetic science still had one more giant to gift to the world whose contribution was to exceed anything that had gone before.

CHAPTER SIX

ISAAC NEWTON AND
THE INVISIBLE BROTHERHOOD

After the collapse of Rosicrucian dreams in Bohemia and Germany and the eruption of the Thirty Years' War that engulfed Europe for a generation, Hermetic hopes for the great reform of society focused on England, which had remained largely uninvolved with the war, if only because Charles I's expedition of the late 1620s had been ignominiously defeated. And when he ran out of funds for another such venture, it was the issue of how to raise money for the army that deeply divided the English.

The ensuing Civil War between the king and Parliament convulsed the country from 1641 until 1649, and ended with the public beheading of Charles I in London and the foundation of Oliver Cromwell's Commonwealth. The years of the Commonwealth and the Protectorate under Cromwell's personal rule, although largely miserable (Christmas was cancelled, for example), were relatively stable.

But before England endured its own upheavals, a number of scholarly refugees who cherished the Rosicrucian dream arrived. England quickly became the repository of the Hermetic reform movement.

The Hermetic tradition had by no means died out in the country. In 1654 John Webster – a Puritan Parliamentary chaplain, astonishingly – wrote a tract proposing that the universities should base their teaching on 'the philosophy of Hermes revived by the Paracelsian school'[1] – in other words, Rosicrucianism. He mentioned the Fraternity of the Rose Cross and strongly recommended John Dee's mathematical works, as well as those of Robert Fludd.

Another important vehicle for the Hermetic tradition in England was a group of philosophers centred on Christ's College, Cambridge, known somewhat misleadingly as the Cambridge Platonists, who were most active in the middle of the seventeenth century. They took the founding philosophy of the Renaissance and blended it with contemporary currents of thought, but at their core was the *philosophia perennis* of Marsilio Ficino – whose heart was Hermetic through and through.[2] One of their most prominent members, Henry More, wrote that his thinking derived from 'the Platonick Writers, Marsilius Ficinus, Plotinus himself, Mercurius Trismegistus and the Mystical Divines'.[3]

Given that list it would be just as accurate, if not more so, to describe this group as the Cambridge Hermeticists, although most historians are content to maintain their bias away from the Hermetica and towards the Greeks. The Cambridge group was in effect the direct continuation of the Florentine Academy of Ficino, the brotherhood of Hermeticists that drove the Renaissance. As historians J. Edward McGuire and Piyo Rattansi demonstrated in the 1960s, the Cambridge Platonists mainly derived their philosophy from the *Corpus Hermeticum* via Ficino and Pico della Mirandola. In a 1973 essay on the Cambridge Platonists, Rattansi wrote that: 'It is now clear that the Neo-Platonism of Ficino and Pico was deeply intertwined with the magical doctrines of the *Corpus Hermeticum* and the later Neo-Platonists.'[4]

The Cambridge Platonists accepted Isaac Casaubon's dating of the Hermetica, but did not acknowledge that this invalidated the philosophy. Henry More regarded only those parts that reflected Christian teaching as 'fraud and corruption in the interests of Christianity',[5] and the rest as genuinely ancient. So, ironically, in More's view, in looking for the original, true theology, the *prisca theologia*, we should pay most attention to those aspects of the Hermetica that are the *least* Christian.

The philosopher regarded as the leader of the group, Ralph Cudworth, while accepting that significant parts of the Hermetica were Christian forgeries, challenged Casaubon's logic. Why did proving some of the Hermetic books to be fraudulent mean that all of them must be? He also argued that if the aim of the forgers had been to build a path into the Church for Egyptian pagans, it would have made more sense to either have adapted genuine books of Hermes or incorporate the major themes of Egyptian thinking into their fakes. So, in Cudworth's view, enough of the underlying philosophy and cosmology remained to draw valid conclusions. And as we will see, his was very close to the current historical position.

THE INVISIBLE COLLEGE

Among the distinguished refugees from the Continent, a key figure was the Polish polymath Samuel Hartlib (1600–62): Hermeticist, Paracelcist, promoter of Dee's mathematical and geometrical works and an astrologer. With his Europe-wide circle of correspondents and contacts he was an 'intelligencer', a sort of one-man clearing house for information. He was a devoted networker in the interests of dissemination of all knowledge, from the intellectually obscure to the political – rather like Gian Vincenzo Pinelli in Padua during Bruno's day.

Hartlib was clearly a Rosicrucian. He worked to found a 'pansophic college' – an institution for the study of all-embracing wisdom, the acquisition of knowledge and its use for the betterment of society. Together with fellow traveller John Amos Comenius (1592–1670), a Czech scholar who also took refuge, briefly, in England, he proposed setting up a Collegium Lucis, or College of Light, for the advancement of learning, but primarily to train up a body of 'teachers of mankind'.[6]

Apart from being influenced by Andreae and the ideal of a learned society working for the advancement of humanity, he took the name for his projected movement, 'Antilia', from Andreae's utopian work *Christianopolis,* which uses the word as a reference to an inner group within his perfect society. Presumably inspired by this was the utopian tale Hartlib wrote, a short pamphlet entitled *A Description of the Famous Kingdome of Macaria* (1641). However, his Rosicrucian connection is made most explicit in his letter he wrote to one of his chief correspondents, John Worthington (1618–71), Master of Jesus College, Cambridge – and one of the Cambridge Platonists:

> The word Antilia I used because of a former society, that was really begun almost to the same purpose a little before the Bohemian wars. It was as it were a tessera of that society, used only by the members thereof. I never desired the interpretation of it. It was interrupted and destroyed by the following Bohemian and German wars.[7]

A tessera is a piece of a mosaic, but as the word was also used in ancient Rome to refer to a ticket, voucher or token, Hartlib seems to be hinting that 'Antilia' was the code name Rosicrucians used to recognize each other. This kind of knowledge implies he was himself a member. Yet another clue lies in the fact that his patron was Elizabeth of Bohemia who, as we have seen, together with her husband was the focus of intense Rosicrucian support.

Try as he might, Hartlib failed to get his projected

pansophic college off the ground, writing despairingly to Worthington in October 1660: 'We were wont to call the desirable Society by the name of Antilia, and sometimes by the name of Macaria, but name and thing is as good as vanished.'[8] Like many other academics and intellectuals who had flourished under the Commonwealth, he had probably simply lost favour at the restoration of the monarchy.

But a month later came the first meeting of what was to become the Royal Society. And it seems that, wherever the initial idea came from, there was an attempt to use it to achieve the 'Antilian' dream.

The train of events that led to the foundation of the Royal Society is more complicated and more esoteric than many modern writers would have us believe. Despite the restrictions of the ongoing Civil War, it began in London in 1645 with an informal meeting of scholars who set out to explore new ideas in natural philosophy – as science was then called. In what was almost certainly no coincidence, the two prime movers were in the retinue of the exiled Charles Louis, Elector Palatine, Frederick and Elizabeth's son. The two were Charles Louis' secretary, Theodore Haak, and his chaplain, John Wilkins. Charles Louis had been invited to live in London by Parliament, whose cause he backed. All very odd for the son of a Stuart – especially given that he was the nephew of the king who Parliament was fighting against.

John Wilkins – the future Bishop of Chester, inventor of the metric system and something of an oddball for a Church of England chaplain – was really the driving force behind the formation of the Royal Society. At the age of forty-two, the highly ambitious Wilkins married Cromwell's sixty-three-year-old widowed sister, presumably a move that did nothing to prevent his inexorable rise. He also wrote a defence of Copernicanism in 1641 (*Discourse Concerning a*

New Planet), and more creatively, a flight of fancy with the self-explanatory title, *The Discovery of a World in the Moone* (1638). His attempt to introduce a new universal language to be used by natural philosophers instead of Latin was terminally halted when his entire print run was lost in the Great Fire of London.

In his hugely popular book *Mathematicall Magick*, published in 1648, Wilkins specifically references the *Fama Fraternitatis*. His book was based – as he freely acknowledged – on mathematical works by Dee and Fludd and even declared that he took the title from Cornelius Agrippa.

It was at this juncture that the now-famous references to an 'Invisible College' appeared. These were in letters written in 1646 and 1647 by one of the most eminent founders of the Royal Society, the chemist Robert Boyle (1627–91) – credited with turning alchemy into chemistry – who alluded to a gathering of scholars and philosophers of which he was a part and which called itself by this mysterious name.

Not only was the intriguing term 'invisible' used in the Rosicrucian manifestos, but it carried clear echoes of the mysterious, even sinister, 'College of the Brothers of the Rose Cross', otherwise known as the 'Invisibles' in Paris. Boyle's comments were almost certainly a kind of Rosicrucian in-joke.

Many writers have seen a connection between this enigmatic group and the founding members of the Royal Society, and hinted at the existence of an anonymous behind-the-scenes cabal. But maybe too much mystery has been read into these connections since the group Boyle refers to is relatively easy to identify. Historian Margery Purver, in *Royal Society: Concept and Creation* (1967), shows that the Invisible College was the circle centred on Hartlib.

The references to the Invisible College appeared in letters

that the young Boyle wrote to Hartlib and make the connection between Hartlib and the activities of the college very explicit. On 8 May 1647 he wrote: 'You interest yourself so much in the Invisible College, and that whole society is so concerned in all the accidents of your life . . .'[9] In other correspondence from around the same time, Boyle calls Hartlib the 'midwife and nurse' of the college.[10]

The Invisible College was Hartlib's Antilia, or more accurately the group of learned men he hoped would become Antilia. Considering this in combination with the 'invisible' clue suggests that it is essentially a Rosicrucian brotherhood. However, this doesn't mean the connection with the Royal Society is nonexistent: Hartlib hovers in the background during its inception and at least initially it embodied his Rosicrucian ideals. And significantly, Boyle was one of the most active founder members.

THE ROYAL SOCIETY
As John Gribbin points out in *The Fellowship* (2005), the Royal Society was the result of two groups coming together. The first was a group that had met informally in John Wilkins' rooms at Wadham College, Oxford, from 1648 and throughout the years of the Commonwealth and which included Boyle and Christopher Wren. The second consisted of royalists with an interest in natural philosophy returning from exile with the Restoration in 1660. The two groups met when attending a series of open lectures at Gresham College in London.

At a meeting on 28 November 1660 a group of twelve natural philosophers and enthusiastic amateurs – including Boyle, Wilkins and Wren and led by William, Viscount Brouncker – decided to form a society for promoting the emerging 'experimental philosophy', or what we now know as the scientific method, using experiment to test hypotheses. They took as their motto *'Nullius in verba'*, literally 'on the word of no one', but 'take no one's word for it' certainly has a more modern ring.

The new society was particularly inspired by the work of Francis Bacon (1561–1626), the English courtier, lawyer and philosopher. His major work is the 1605 book *The Advancement of Learning*, which, as its title suggests, surveyed the state of scholarship in his day and proposed ways in which natural philosophers might extend their knowledge. He argued for a methodical and systematically organized approach to investigating the natural world, and also called for a united international 'fraternity in learning and illumination'.[11]

Historians long regarded Bacon as the archetypal voice of reason, a beacon of light in an age of superstition, but in 1957 the Italian historian Paolo Rossi's *Francis Bacon: From Magic to Science* challenged this long-held view. From closely examining Bacon's life and work, Rossi showed that he was as much a devotee of the Hermetic tradition as the other thinkers we have so far discussed. Rossi notes in particular the 'influence of the hermetic doctrine' on Bacon's ideas on the nature of metals.[12] He also, according to Rossi, firmly believed in the *anima mundi*. Basically the great man was another passionate disciple of the Renaissance occult philosophy (although he wanted to reform that, too). He included natural magic, astrology and, particularly, alchemy, within his fields of knowledge. He was just careful not to draw attention to them.

Ernest Lee Tuveson observed that Bacon's 'conception of natural processes owes much to hermeticism, and other traditional [i.e. esoteric] sources',[13] and asked why he therefore condemns the likes of Dee, Fludd and Paracelsus. He concludes that, although Bacon shared their underlying philosophy, he disagreed about the methods that should be used to put it into practice, advocating the application of objective reasoning instead of magic. However, we can suggest a rather more expedient, if not cynical, motive: Bacon was in need of the king's favour, and was all too aware that there were certain subjects that were best avoided.

King James I, offspring of the doomed Mary, Queen of Scots was a weird little man with a paranoid terror of witches and would go to any lengths to protect himself from the threat of witchcraft, real or imagined (mostly the latter, but your innocence would hardly matter if you were accused and condemned and rolled down a hill inside flaming barrels on his orders). It was James' horror of all things occult that had been responsible for Dr Dee's decline.

In many ways Bacon was Dee's successor, another man of many talents who was involved in court and diplomatic activity under the patronage of the monarch. He rose to prominence at court immediately after Dee's fall from grace, for example producing the masque performed on the day following the wedding of Princess Elizabeth and Frederick V of the Palatine. But rising to prominence in those days was no guarantee of a long happy life – one had to work at it constantly, which usually involved shameless amounts of regal boot-licking.

Bacon was a highly ambitious man. As Arthur Johnston notes in his introduction to a 1973 edition of *The Advancement of Learning*, Bacon's life was 'a long pursuit of political power'.[14] In practice this meant mounting a campaign to attract the king's attention and favour – which certainly worked. In fact, as Jerome R. Ravetz, lecturer in the History and Philosophy of Science at the University of Leeds cautions: 'All Bacon's published writings are propaganda; their function was to convert his audience, and their relation to his own private views was purely incidental.'[15]

Bacon enjoyed a succession of appointments at court that culminated in his elevation as Lord Chancellor in 1618. As an appeal to James I's intellectual pretensions, *The Advancement of Learning* opened his campaign of self-advancement and eventually earned him a job putting his proposed reforms of learning and education into practice. Fittingly, the very first paragraph includes the hardly subtle appeal for 'the good pleasure of your Majesty's employ-ments'.[16] The book was addressed directly to James, whom he overtly flatters: 'There hath not been since Christ's time any King or temporal Monarch, which has been so learned in all literature and erudition, divine and human.'[17] Bacon certainly knew how to lay it on with a trowel, echoing Bruno's wildly over-the-top flattery of James' predecessor, Elizabeth I. More interestingly he dared to liken James to Hermes Trismegistus:

> Your Majesty standeth invested of that triplicity, which in great veneration was attributed to the ancient Hermes; the power and fortune of a king, the knowl-edge and illumination of a priest, and the learning and universality of a philosopher.[18]

This particular description of Hermes is taken from Marsilio Ficino – which presumably Bacon relied on James not knowing.

Bacon's call for a 'fraternity in learning and illumination' may have influenced the authors of the Rosicrucian manifestos, but if so he was also influenced in turn by them. There are clear signs that he was familiar with the *Fama Fraternitatis* in his utopian *New Atlantis*, published in 1627, the year after he died, and which was a particular influence on the Royal Society's founders. Bacon seems also to have read and digested Campanella's *City of the Sun* (published four years earlier) – or maybe it is a coincidence that his plot, too, involves shipwrecked sailors encountering the inhabitants of a perfect society (the preservers of an early, pure form of Christianity, whose officials wear white turbans bearing red crosses)?

Given Bacon's unofficial interests, it is rather ironic that he is seen to represent the beginning of the divergence of magic and science.

A more elusive and unequivocally arcane influence on the origins of the Royal Society was Freemasonry. Although the origins of Freemasonry are still controversial and obscure, whatever its roots it had certainly emerged as a significant institution by the mid-seventeenth century. Many historians have seen the Brotherhood as a repository of the Hermetic tradition, though this is not to suggest that Freemasonry is only about Hermeticism.[19]

Significantly, Masonic writer Robert Lomas points out that one of the rituals an initiate undergoes when entering the second degree makes specific reference to the heliocentric theory: 'The sun being at the centre and the Earth revolving around the same on its own axis . . . the sun is always at the meridian with respect to Freemasonry.'[20] Even in the mid-seventeenth century heliocentricity was still not fully accepted – and in Catholic countries was an outright heresy – so the Masons' emphasis is all the more telling.

A Masonic influence on the early Royal Society is now generally acknowledged, but its extent and significance are more controversial. However, what is less contested is that the Society's main connection with the Freemasons was embodied by one of the driving forces behind its foundation – the man who secured its royal patronage, Sir Robert Moray (1609–74). His Masonic initiation in 1641 has the distinction of being the first to be recorded on English soil.

Described by Lomas as 'a first-rate fixer and born survivor',[21] Moray was a mixture of James Bond and soldier of fortune – but with mystical trappings. His origins remain obscure, but he first made his mark as a member of the Scots Guard of Louis XIII's army in 1633, when he spied for Cardinal Richelieu. He then turns up as the quartermaster of the Scottish Covenanters' Army when it marched on England in 1640. This campaign was part of the struggle over control of the Church in Scotland during which the Scots occupied parts of northern England including Newcastle, where Moray was initiated into a Masonic lodge on 20 May 1641. It is generally thought that he used his Masonic connections for intelligence work. After the end of the Covenanters' campaign, he returned to the French court for yet more soldiering and spying and eventually established himself as an emissary between the French court and that of Charles I, who knighted him in 1643. Moray went on to become the King's secretary, and after Charles' execution he joined Charles II's exiled court in Paris and became heavily involved in the negotiations to set him on the restored English throne.

With the monarchy restored, Moray based himself in London, where he became one of the twelve that formed the nucleus of the Royal Society. At their second meeting in December 1660, he took the encouraging news that the King approved of their aims and was prepared to give the society his royal endorsement.

However, all was not well within the ranks of the early Society. It is evident that there was a struggle behind the scenes between those who followed a more Hermetic/ Rosicrucian model of a learned society and those who shared Bacon's vision. The Hermetic version lost. This happened during the securing of the royal charter, which is normally portrayed as a simple intervention by Sir Robert Moray, enthusiast for the project and close friend of the King. But papers lost for three hundred years and rediscovered in the mid-twentieth century reveal a welter of plotting behind the scenes. Prime mover in this was Baron

Skytte, a Swedish nobleman and confidant of King Karl Gustav, who was in London to promote the creation of a Protestant Alliance. Also interested in the new learning, Skytte attended the lectures at Gresham College.

On 17 December 1660, Hartlib wrote to John Worthington that since his last letter of ten days before:

> I have recd some other papers, that have been confided to me, holding forth almost the same things as the other Antilia (for be not offended if I continue to use this mystical word) but, as I hope, to better purpose.[22]

These papers, he goes on, were sent to him by Skytte, and comprise:

> . . . the propositions which were made to his Majesty by the Lord Skytte, and . . . a draught for the royal grant or patent wch is desired for the establishment of this foundation. Thus much is certain, that there is a meeting every week of the prime virtuosi, not only at Gresham College, in term time, but also out of it . . . They desired that his Maj leave that they might thus meet or assemble ymselves at all times, wch is certainly granted. Mr Boyle, Dr Wilkins, Sr Paul Neale, Viscount Brouncker are some of the members.[23]

Skytte had evidently resurrected Hartlib's plans. However, although Boyle supported them, they ultimately failed because of opposition from other founding members. Hartlib wrote to Worthington in April 1661, 'There becomes nothing of Lord Skytte's business, & I believe the other virtuosi will not have it that it should go forward.'[24] After the royal charter was granted in July 1662, Skytte returned to Sweden, and Hartlib died the following year.

In response to the Royal Society's publication in 1667 of

its early official history, by Thomas Sprat – later Charles II's chaplain and the Bishop of Rochester – Worthington railed that the society was 'materialistic and for nothing but what gratifies externall sense.'[25] His outburst underlines criticism that the Society had failed to realize its full potential because it had rejected the more philosophical and metaphysical elements championed by Hartlib and Baron Skytte. The essential difference between the two visions of a learned society is that Hartlib's had the reforming aspect that went back to Andreae and the Rosicrucian manifestos, and beyond that, to Campanella – and ultimately to Bruno.

One wonders exactly why a society, no matter how well connected, would be in quite so much of a hurry to rush out its official history, just seven years after it was founded. Their haste may represent a desire for the victors to etch their triumph in the minds of its readers, but it also suggests the promotion of a version of events that was economical with the truth.

Another sign of the Royal Society's Hermetic eclipse was the sidelining of John Wilkins, the man who had started the club at Wadham College and a Rosicrucian-friendly founder. Although he was appointed as the Society's secretary, he shared this role with a newcomer, the German-born theologian and diplomat Henry Oldenburg, and was soon marginalized.

Was the struggle over the direction and control of the Royal Society simply about the scientific philosophy it should adopt? In fact there appears to have been more to it even than that. One result of the organization of the new Society was that Oldenburg, as its foreign secretary, inherited Hartlib's network of correspondents, and he undoubtedly used his position for intelligence-gathering of a more politically sensitive kind.

Robert Hooke, the Royal Society's curator responsible for organizing experiments, complained that Oldenburg 'made a trade of intelligence'.[26] In fact, he used his network for gathering not just scientific but also political information, the latter on behalf of the Secretary of State, Lord Arlington, even arranging for all the Society's correspondence from abroad to be delivered to the office of Arlington's under-secretary. Oldenburg was imprisoned in the Tower of London for two months as a suspected spy during the Anglo-Dutch war of 1667, only being released when peace was made.[27]

As Sir Robert Moray was also a spy, this raises the question of whether one reason the Royal Society was created was as a cover for intelligence-gathering. After all, it still remains unclear why Charles II was quite so interested in the Royal Society. What was in it for him? Although this suggestion might seem absurd, bear in mind that in its early days the Society was not the celebrated and distinguished institution it is today. It was only when Isaac Newton became its president in 1703 (his presidency lasted for twenty-four years) that it could bask in his immense prestige. John Gribbin writes that by the end of Newton's tenure the society has completed 'the process whereby a gentleman's club became a pillar of the establishment'.[28]

If the reforming side of the Hermetic tradition had been extinguished by the time the Royal Society came into being, its influence over the scientific revolution had not waned. And it reached its final, and greatest, flowering in the person of 'the most outstanding scientific intellect of all time',[29] Isaac Newton.

THE GREATEST SCIENTIFIC GENIUS
Isaac Newton (1642–1727) is widely regarded as the greatest scientific genius in history, and his masterwork, *Mathematical Principles of Natural Philosophy* (*Philosophiae naturalis principia mathemetica*) – usually known, reveren-

tially, simply as the *Principia* – is deemed the single most influential book ever written. His elucidation of the laws of motion and of gravity effectively *created* the modern world: mechanics and most forms of transport, including space travel, would be impossible without them. Newton even created the mathematical system, infinitesimal calculus, needed for his work – in itself no small achievement. After all, this and most other books would never see the light of day if writers had first to invent laptops – or, more appositely, writing itself. But Newton had the vision to know what he needed to be great, then went ahead and made it all happen with the unswerving, if often anti-social dedication of genius.

Despite being from a humble background, Newton managed to rise to fame and fortune. He was the only child of a Lincolnshire farmer who had died by the time he was born on the farm at Woolsthorpe near Grantham, in the first year of the Civil War, on Christmas day 1642. He was a sickly child, and for his whole life he would be a solitary soul. From the age of three Newton was brought up by his grandmother, his mother having married the rector of a nearby parish. He hated her and his stepfather for abandoning him and went so far as to threaten to set fire to their house with both of them in it, but the rector's relative wealth would in the end prove useful to him.

From the beginning, Newton was fascinated by mechanics and delighted in making machines such as a mini mouse-powered windmill. He was entranced by how things worked. A life-changing moment came at the county fair when he bought a prism from an itinerant salesman, which stimulated his obsession with the phenomena of optics. Naturally, he was expected to be a farmer like his father, but when he was twelve an understanding uncle – a Cambridge graduate – realized that was not his destiny and secured him a place in a school at Grantham, where he also had to work as a servant to wealthier students.

John Gribbin describes one of Newton's early practical jokes:

> He . . . caused one of the earliest recorded UFO scares by flying a kite at night with a paper lantern attached to it, thereby causing 'not a little discourse on market days, among the country people, when over their mugs of ale'.[30]

Newton was also not afraid of experimenting on himself. On one occasion he stuck a bodkin behind his eye to test its effect on his eyesight. On another he stared at the sun until he almost went blind – mercifully the effects were only temporary. Some might think he carried being a genius to a ludicrous degree.

Newton won a scholarship to Trinity College, Cambridge, in 1661, where he seemed merely an average student. Little from his time at Cambridge suggested the historic genius he would become. In 1665, just after he graduated, the college closed because of the plague that was sweeping the country, and he returned to Woolsthorpe for two years. What was a disaster for so many actually ended up being the making of Newton. It was at Woolsthorpe that he experimented with the prism, unravelling the secrets of light. It was also there that he devised the calculus, which he termed 'fluxions'.

And momentously, it was also at Woolsthorpe that he first began to think about the problem of gravity. The story of the falling apple stimulating his thinking of about gravity was Newton's own. The apple tree is still there – the original was cut down long ago but a new one grew from the stump. He realized that whatever caused apples to fall also kept the Moon in its place and determined and governed the motions of the other planets, and therefore the Earth. It would take him twenty years and a radical shift in his thinking to refine and build on his original intuition.

Once the plague was over, Newton returned to Cambridge as a Fellow, and became professor of mathematics in 1669, at the age of twenty-seven. Immediately this caused a problem. At that time newly elected Fellows had to be ordained priests (Anglican, of course) although Newton argued – ultimately to Charles II, who had to approve the appointment – that he should be exempt from this rule.

Although Newton was deeply spiritual, he kept his beliefs so private that even today no one is certain what they were. But the very fact he was so circumspect – and had challenged the ordination rule – suggests that his beliefs were at odds with the dogma of the Church of England. Newton certainly seems to have been a Christian but of a heretical kind, although there is no consensus about its exact nature. Ironically for a Fellow of Trinity College, he definitely doubted the doctrine of the Trinity, as he wrote a book about it that he wisely decided not to publish. He seems to have doubted that God and Christ were 'of one substance', and may even have regarded Jesus as non-divine. He refused the sacrament on his deathbed.

Newton first attracted the attention of his peers through his pioneering work on optics and light, for example inventing the first practical reflecting telescope, using a mirror instead of a lens. As a result, he was elected a Fellow of the Royal Society in 1671. It was at the end of the decade that he returned in earnest to his research into gravity, prompted by a dispute with Robert Hooke.

Newton poured all his thoughts and the results of his experimental work into his monumental achievement, the *Principia*, begun in 1684 and published three years later. The full title of the *Principia* was itself revolutionary, since it declared that natural philosophy was explicable and expressible in mathematical form. Astronomers such as Copernicus and Kepler had used mathematics and geometry, and Galileo had taken their application a step further, but to Newton mathematics was at the very heart of science.

One consequence of the *Principia* was the final proof of Copernicus' theory. Newton demonstrated that his theory of universal gravity accounted for Kepler's laws of planetary motion, which was in turn derived from Copernicus' heliocentric model. This was the great watershed in the history of cosmology: after the *Principia* was published, it was impossible to doubt the heliocentric theory. To Bruno, of course, this would have represented only a partial success. Global acceptance of heliocentricity was due to usher in a golden age of universal Hermeticism, after all. But things had changed since Bruno's day . . .

The *Principia* was an immediate sensation, although rather like Stephen Hawking's *A Brief History of Time*, it was 'one of the least-read bestsellers of the age'.[31] After it was published Newton moved to London, where he became a celebrity, albeit a rather reclusive and curmudgeonly one. He was knighted in 1705 by Queen Anne – the first 'scientist' to be honoured in this way. Both she and her successor George I would heap great honours on him. Newton became Warden of the Mint in 1696, then Master of the Bank of England, and was elected President of the Royal Society in 1703, a position he retained until his death. He was a Member of Parliament for two short periods. When he died in 1727 it was a cause for national mourning. In honour of the occasion of his state funeral in Westminster Abbey, the poet Alexander Pope penned the famous lines:

Nature, and Nature's Laws lay hid in night:
God said 'Let Newton be!' and all was light.

However, Newton was anything but the sort of materialist-rationalist so prevalent today among the ranks of scientists, who believe all spirituality is a form of superstition. It is now well known that Newton's major preoccupation was not gravity or the laws of motion or optics, but alchemy. The first biography that mentioned this was in 1855 but even after that it was a subject that was glossed over fleetingly and apologetically. More recently, however, historians of science have begun to acknowledge that Newton's esoteric interests did not only play a vital part in his thought processes, but also actively assisted him in making his great discoveries.

Richard S. Westfall, Professor of the History of Science at Indiana University and author of a major biography of Isaac Newton, wrote in 1972:

> One lively and active facet of the lively and active enterprise that is Newtonian scholarship today is the continuing revelation of the presence in Newton's mind of modes of thought long deemed antithetical to the modern scientific mind.[32]

One of the first to realize the importance of Newton's esoteric side was John Maynard Keynes, the leading twentieth-century economist and great collector of Newton's alchemical writings, who in a paper read to the Royal Society in 1946 commented that 'Newton was not the first of the age of reason. He was the last of the magicians . . . '[33] He went on (his emphasis):

> Why do I call him a magician? Because he looked on the whole universe and all that is in it *as a riddle,* as a secret which could be read by applying pure thought to certain evidence, certain mystic clues which God had laid about the world to allow a sort

of philosopher's treasure hunt to the esoteric brotherhood.[34]

On Newton's death, 169 books on alchemy were found in his personal library – making up one-third of his collection. In fact, it transpires from all his writings that his main esoteric preoccupation was the quest for the philosopher's stone, and he was particularly fascinated by the work of the French alchemist Nicolas Flamel (*c.* 1330–1418).

Most of Newton's alchemical papers – of which he produced a vast number, over a million words – collected by Keynes and others, are now in Jerusalem, in the Jewish National Library. As befits the work of a genius with a need to be secretive, they are written in elaborate codes, and many of them have yet to be deciphered.

Alchemy was against the law, and could even attract the death penalty (although in a curious excess of official spite, alchemists were to be hanged on gilded scaffolds adorned with tinsel, so at least their demise was pretty in a trashy sort of way). Legal disapproval existed not for reasons of religious intolerance, or because alchemy was considered fraudulent, but because of the fear that alchemists might succeed in making gold, and thereby undermine the economy. So it is an exquisite irony that the Establishment saw nothing wrong in putting Newton – an alchemist to his gilded fingertips – in charge of the Bank of England and of the Royal Mint, even entrusting him with the re-minting of the entire currency in the 1690s.

Like many esotericists before and after him, Newton was a great believer that the earliest civilizations, such as Egypt, knew more than people in his own day – that they possessed the *prisca sapientia,* or 'ancient wisdom'. He was in no doubt that the Greeks had learned everything they knew from the Egyptians. He also believed that the Bible was one of the sources of the ancient wisdom, and that it contained prophecies relevant to his own time, particularly in the Book of Revelation. Besides studying many other ancient temples and buildings, he was fascinated by the Temple of Solomon, and devoted considerable energy to the study of its design, dimensions and proportions, which he believed incorporated ancient truths.

Like many thinking people of the post-Renaissance world, Newton was also particularly interested in Rosicrucianism, possessing copies of the English translation of the manifestos and Michael Maier's works, which he annotated heavily. In his copy of the English translation – now held in Yale University Library – he wrote a lengthy note on the purported history of the Fraternity of the Rose Cross. Referencing Maier in particular, the note ends, 'This was the history of that imposture.'[35] This quote is often cited as evidence that Newton rejected Rosicrucianism. However, it actually refers only to the Christian Rosenkreutz legend in the *Fama*, which Newton recognized as an allegory or *ludibrium*.

The source of Newton's obsession with the esoteric is particularly illuminating. He undoubtedly started out as a mechanist, pure and simple, reserving a special admiration for Descartes. However, in the mid-1670s he changed radically, embracing a far more arcane worldview. The reason for this can be traced back to the influence of the Cambridge Platonists, especially that of Henry More, who – nearly thirty years Newton's senior – was an old boy of the same school in Grantham. As we saw earlier, this woefully misnamed group were fundamentally Hermeticists, part of an unbroken line of a spiritual brotherhood stretching back to Marsilio Ficino, who rediscovered the works of Hermes Trismegistus. At least one member of the Cambridge Platonists, John Worthington, was also part of Hartlib's Invisible College, itself a direct continuation of the Rosicrucian Antilia, which was intimately connected to Bruno's reforming campaign and the Giordanisti.

One of the first papers to recognize the importance of Newton's Hermeticism was by J. Edward McGuire and Piyo M. Rattansi, both lecturers in the history and philosophy of science at Leeds University. Published in the *Notes and Records of the Royal Society* in December 1966, the paper, 'Newton and the "Pipes of Pan"', was based on a study of Newton's draft of rewritten sections of the *Principia*, which he wrote in the 1690s for a proposed new edition that was to have included more on the esoteric. McGuire and Rattansi explore the influence of the Cambridge Platonists on Newton's thinking, concluding that:

> In re-examining Newton's relation to the Cambridge Platonists, we shall see that he did not merely borrow ideas from them, but was engaged in a private dialogue whose terms were set by a certain intellectual tradition.[36]

But which 'certain intellectual tradition'? They go on to identify it as the 'most elaborately developed Renaissance *prisca* doctrine' found in the works of Ficino and Pico, which were derived from the *Corpus Hermeticum*.[37] McGuire and Rattansi add that 'Newton, and the Cambridge Platonists, saw their task as the unification and restoration of this philosophy.'[38] In the words of Richard Westfall, as a result of Newton's association with the Cambridge Platonists, 'the Hermetic influence bade fair to dominate his picture of nature at the expense of the mechanical.'[39]

Newton frequently cited Hermes Trismegistus in his alchemical and esoteric private writings and wrote a detailed commentary on the *Emerald Tablet* (which was considerably longer than the original). An American historian who specialized in Newton's alchemy, Betty Jo

Teeter Dobbs, comments on the extent of Newton's passion for Hermes explaining that 'Newton's study of Hermes Trismegistus extended over a period of at least twenty years, possibly longer.'[40]

Newton's Hermeticism transformed his thought in precisely the opposite direction to that which we have come to expect in the twenty-first century. The modern perspective is that people started with vague and supernatural explanations for how things work, but eventually came to understand them in purely mechanical and logical terms. But Newton moved from mechanics to magic. As Westfall writes:

> In Newton, peculiarly Hermetic notions fostered the crucial development of his scientific thought, and in the concept of force became a central element both in the enduring science of mechanics and the accepted ideas of nature. The fundamental question for Newtonian scholarship, as it appears to me, is not the presence of Hermetic elements in his philosophy of nature; their presence has been demonstrated beyond reasonable doubt. The fundamental question is the mutual interaction of the two traditions in the development of Newton's scientific thought.[41]

It is now recognized that it was not an apple falling on Newton's head – or even less dramatically simply plumping to the ground in front of him – which gave him his eureka moment, but delving into the pages of the Hermetica. And it bestowed on him nothing less than the key to unlock the mysteries of nature.

It is not simply a matter of Newton hitting on the physical laws of nature by drawing analogies with the Hermetic principles. He *applied* those principles to physical systems. For example, the big resistance to his explanation of gravity was that many considered it to be too 'occult'. His notion of gravity as a force that acts across space, at a distance, and does so in the way it does purely as a consequence of the nature of the universe, was drawn straight from the magical laws of sympathy and attraction as expounded in the Hermetica. (Newton put it more succinctly, declaring 'Gravity is God'.) The law of gravity invokes principles relating to forces that act between the Earth and heavenly bodies that feature – in very different language, of course – in *Asclepius*, the same work that inspired Copernicus.

And Newton's certainty that the heliocentric model was correct also seems primarily to have been drawn from his knowledge of the Hermetica, rather than from the works of Copernicus or Kepler. In a discussion of the mysteries of ancient Egypt he wrote:

> It was the most ancient opinion that the planets revolved about the sun, that the earth, as one of the planets, described an annual course about the sun, while by a diurnal motion it turned on its axis, and that the sun remained at rest.[42]

Of course, the obvious source for this understanding of the Egyptians is, once again, *Asclepius* and the other Hermetic texts.

While most scholars recognized Newton's *Principia* as a work of genius, a sizeable number immediately dismissed it as a farrago of occultism. Richard S. Westfall comments:

> The cry of occult qualities greeted the publication of the *Principia*. In more than one sense, the mechanists who raised the cry were justified. Not only did the concept of attraction violate their sense of philosophic propriety, but the origin of the concept was the very

Hermetic tradition they suspected . . . The champions of mechanical orthodoxy failed to realise what benefit the Hermetic idea could bestow on the mechanical philosophy of nature.[43]

Of course nobody today would dare side with Newton's contemporary detractors. Newton's genius is now universally recognized. And yet there are still those who can't see the significance of the esoteric facet of his life and work. If nothing else, his modern critics show themselves on this major point to be giants of condescension and pygmies of understanding.

In his *God is Not Great* (2007), Christopher Hitchens unhesitatingly describes Newton as 'a spiritualist and alchemist of a particularly laughable kind'.[44] Apparently in today's era of education and enlightenment even your average journalist and literary critic possesses a greater intellect than poor befuddled old Isaac Newton. But the reality is simple: if Newton had never had become privy to the Hermetic philosophy, he would never have achieved his work and the world would be – literally – much the poorer for it. It is universally acknowledged that if the *Principia* had never been written, our modern technological world would not exist. But without the Hermetica, Newton would never have written the *Principia*. Emphatically Newton did not make his great scientific discoveries *despite* his esoteric beliefs, but *because* of them.

The same is true of Copernicus, Kepler, Gilbert, Galileo, Kircher and Leibniz. All of these great scientific minds either drew their inspiration directly from the Hermetica or indirectly from the works of other Hermetic masters – usually Bruno. Without that extraordinary philosophy and its accompanying curiosity, they would never have realized that mere men could be giants, gods of thought to whom anything was possible and that freedom from the tyranny and poverty of intellect that marked the reign of the Church of Rome was, indeed, possible.

This raises some other important questions: If the Hermetica was this wondrous intellectual instrument, where did it originate? How did its authors come to know such secrets? Who were *they*? And was Newton right? Did the Hermetic texts embody the greatest wisdom of ancient Egypt?

CHAPTER SEVEN

EGYPT'S TRUE LEGACY

The mysterious collection of works known as the Hermetica may have illuminated the path for many of the world's greatest scientists and philosophers, who believed it to be the authentic repository of ancient Egyptian wisdom, but in 1614 Isaac Casaubon threatened to completely undermine their position, declaring authoritatively that the books were 'only' about a millennium and a half old, dating from the early centuries CE. Modern historians agree that Casaubon, who employed philological techniques (the analysis of language and literary style), reached roughly the right conclusions, even if for the wrong reasons, at least as far as the actual composition of the Hermetica is concerned. Its sources, however, are quite another story.

As we have seen, Casaubon demonstrated that the Greek of the Hermetica does not belong to the classical period but is a later style altogether, which dates from the late centuries BCE and early centuries CE. This timeframe makes sense, as this was when Egypt was ruled successively by the Greeks and the Romans, a period beginning in the 330s BCE during Alexander the Great's most feverish bout of empire building.

After Alexander's death his general Ptolemy declared himself pharaoh, establishing the Ptolemaic dynasty that lasted for three centuries until the death of Cleopatra.

During this time Hellenic customs, lifestyle and language took hold, at least among the top strata of Egyptian society. In 30 BCE, after the second most famous snake in history (after the chatty tempter that appears in Genesis) had done its worst to the Queen of the Nile, the Romans took over, although Egyptian-born Greeks continued to be over-represented among officialdom. Greek, rather than Latin, remained the *lingua franca* of the eastern half of the Roman Empire.

This means that the Hermetic texts were composed at some point between the beginning of the Greek domination and their first mention in Christian works in the third century CE, a period that lasted around 500 years. This may not pinpoint the precise historical moment of the Hermetica, but it still places them well after the golden age of the Egyptian civilization. So how could they contain the secrets of the pyramid builders?

This question highlights a flaw in Casaubon's argument. Establishing that the Hermetic books dated from the period of Greek and Roman domination was hardly earth-shattering. If they had been composed any earlier they wouldn't have been written in Greek of any style, but in Egyptian. And of course the fact that they were composed in Greek does not necessarily mean that the ideas they expressed were conceived at that time. They could, for example, have been written to explain an Egyptian belief system to Greek-speakers, or just as easily have comprised a translation of Egyptian wisdom texts. These fairly obvious objections didn't escape the seventeenth-century Cambridge Platonists, who used similar arguments against Casaubon.

That is why the second part of Casaubon's case, which he based on the content of the Hermetica rather than the style, was important. By demonstrating that certain sections had been influenced by concepts from Plato and the New Testament – particularly John's Gospel and some of Paul's letters – Casaubon believed he had proved that the texts were composed from scratch after the time of Christ.

Modern historians have roundly rejected this part of Casaubon's argument, seeing no direct connection at all between the New Testament and the Hermetica. Any potential link is indirect, as both texts derive from the same blend of theological and philosophical speculation, drawn from various cultures including the Hellenic, Iranian, Judaic – and, of course, the Egyptian – which were being explored at that time.

As we have seen, what really excited Renaissance Hermeticists was the parallel between the description of God's Word in the *Pimander* and the Word/Logos passage that opens John's Gospel. However, the unknown writer of this gospel took the concept from the work of Philo of Alexandria (*c.*20 BCE–*c.*50 CE), a Hellenized Jew who blended Jewish theology with ideas from the great intellectual melting pot that was his own city. The Hermetica also drew from the same pool of ideas, so any connection between the Word in *Pimander* and the Gospel of John is indirect. It doesn't even necessarily mean that the Hermetica came after Philo, since the ideas he drew on had been in the philosophical mix for some time. And – as we are about to see – part of this included home-grown Egyptian traditions, which almost certainly provided the inspiration for the Hermetic description of God's Word. Although it is only too easy to pity Casaubon, there was simply not enough information available in his day to make a proper analysis.

So now we're back where we started. As was believed before Casaubon put the feline among the feathered creatures, the Hermetic books may have contained traditions, not to say secrets, from the old Egypt, the Egypt untainted by the trendy Hellenic glamour of its occupiers. So is it possible to deduce when the Hermetica were written, and by whom? And, more importantly, what were their sources? Was Hermeticism invented in Greek or Roman Egypt, or did it draw on older traditions?

THE ORIGINAL TIME LORD

During the eras of Greek and Roman rule, Egypt – and particularly Alexandria with its famous library – was the crucible where the intellect's gems of the known world came together. As well as native Egyptians, those of Greek descent and peoples from all over the Empire, the great seaport also boasted large Jewish and Samaritan communities. Trade routes brought Iranians, Arabs and even Indians to the city, carrying their traditions with them.

Even so, and despite the flaws in Casaubon's work, for a long time historians still assumed that the philosophy and cosmology found in the Hermetica were derived from Greece. It just had to be Greece – after all, weren't the best things always Hellenic? Positively pickled in the classics, the academic world refused to dip a toe into any other culture. But over the years this became increasingly untenable, and with scholarly huffing and puffing, beard stroking and dragging of feet, it was gradually acknowledged that native Egyptian thought must have had at least a supporting role in shaping the Hermetic books.

Doubts about the purely Greek origin of the Hermetica began to surface in the early twentieth century, when university men and women realized key elements of its philosophy and cosmology could not be attributed to any identifiable Greek source. But there was controversy about where they did come from, the main candidates being native Egyptian, Judaic and Iranian traditions.

Perhaps understandably, at first it was mostly Egyptologists who held out for a home-grown influence. Then in 1904 Richard Reitzenstein, the eminent German scholar of Gnosticism and the Hellenic religions, made the groundbreaking suggestion that the Hermetica was the product of a religious community in Egypt. (He did, however, later change his mind, looking towards Iran instead.) From the mid-twentieth century many scholars – particularly in France – joined the pro-Egypt camp. It gradually became a question of not *if* there was an Egyptian influence, but of its true extent. A consensus also emerged that at least the core parts of the Hermetica dated from the early years of Greek domination, rather than towards the end of the era, as Casaubon came to believe.

Key scholars in this process were the French historian Jean-Pierre Mahé and, more recently, Garth Fowden, the British professor of antiquity who is currently Research Director at the Institute for Greek and Roman Antiquity in Athens. As the title of his 1986 book *The Egyptian Hermes* suggests, he amassed the considerable evidence of a strong home-grown, Egyptian, influence on the Hermetica.

Although presented as a characteristic Greek dialogue, the Hermetic texts don't quite fit that genre. Instead of presenting a discussion between philosophers, as in the Greek tradition, the texts present a question-and-answer session between master and pupil – which is more in keeping with the traditional Egyptian wisdom literature.[1] The Hermetic texts are therefore a kind of stylistic hybrid of the Egyptian and Hellenic forms. Maybe the writers were consciously striving to make their work more Greek-friendly.

The books are obviously the product of different writers – which accounts for their inconsistencies – although they belonged to the same school or cult. All the authors here are anonymous, simply attributing their works to Hermes, a typically Egyptian practice.[2] This was quite different from the Greeks or Romans, who routinely hyped up their celebrity philosophers without making any claims of divine authorship. This is another important indication that, while written in Greek, the mindset behind the Hermetica was authentically Egyptian.

Another clue comes from the astrology and astronomy described in the Hermetica. The Egyptians divided the night sky into thirty-six parts or decans, each linked to a prominent constellation or star. During the Greek period, the more familiar twelve-sign zodiac took over, but the astronomy described in the Hermetica sticks to the thirty-six-decan system, so at least the origin of the Hermetica in this one major respect was truly Egyptian.[3]

A more important clue comes from the attribution to Hermes, the Greek deity who was always associated with the Egyptian wisdom-god, Djehuty, or Thoth in its Greek rendering. He ruled over learning and was the inventor of writing and the calendar and 'keeper of the divine words',[4] hence his titles, 'Lord of Time' and 'Reckoner of Years'.[5] A minor function was his association with healing: he was, for example, credited with inventing the enema.

Hermes and Thoth are by no means direct parallels, though. In the Greek pantheon, Hermes was the patron deity of many aspects of life, but not of knowledge and learning. He was the god of cunning and cleverness, but that isn't the same thing. It is thought that the association developed as a result of Hermes' more significant role as guide of dead souls, which oddly echoes Thoth's rather secondary job as helper and guide of the deceased, specifically the dead Osiris.[6] The telling fact is that the characteristics of Hermes Trismegistus as portrayed in the Hermetica are more in line with those of Thoth, not the Greek Hermes, strongly suggesting that the cult or school behind the Hermetica was Egyptian.

Then there is the famous epithet for Trismegistus, 'Thrice Great', which only makes sense as a Greek translation of a typical Egyptian honorific. For emphasis Egyptians repeated the glyph for 'great', literally saying 'great great'.

But in cases of truly mind-blowing greatness, they would use it three times, as in the all-important 'great great great Thoth'. The most natural Greek translation would be 'three times great'. More significantly, this practice seems to have been specifically reserved for Thoth, which seems to be the origin of 'Thrice Great Hermes'.

In 1965 an inscription was found dating from around 160 BCE, written by a priestly scribe named Hor (Horus). Inscribed in the late form of Egyptian script known as Demotic, it appeals to 'Isis, the great goddess, and Thoth, the three times great'[7] – the last phrase simply repeating the Demotic character for 'great' three times. This is the earliest known use of this form of address. In his account of this inscription, Egyptologist J. D. Ray makes the following highly pertinent observation:

> It is not the point of least interest in our document that they should provide the earliest clue to the origins of a most remarkable figure in the history of thought, a philosopher, whose reputation as the sage 'Trismegistus' was transmitted through the Middle Ages and Renaissance to influence even such fore-runners of modern thinking as Bruno and Copernicus.[8]

In fact, the mindset behind the Hermetica as a whole is Egyptian. The authors 'think in terms of a whole milieu populated with ancient Egyptian gods and sages'.[9]

The other characters in the dialogues are a mix of Egyptian – including Isis and Thoth himself, who appears under the name of Tat – and Greek. But even the Greek characters have specifically Egyptian associations. Agathodaimon (or Agathos Daimon), a minor god in the Greek pantheon, became patron deity of Alexandria, where he was associated with Osiris and his Hellenised semi-alter ego, Serapis. More central to the Hermetica is the character of Asclepius, a supposed descendant of the Greek healer-god, who appears in several books. But even here there is an important clue to the origins of the Hermetica. The Greeks identified Asclepius with the Egyptian god of healing and medicine, Imhotep, who was a rare example of Egyptian deification of a living person.[10] In *Asclepius*, Hermes declares that the eponymous main character's illustrious ancestor was a man who became a god. Imhotep was vizier to the pharaoh Zoser and architect of the first of the great pyramids, the Step Pyramid of Saqqara, built around 2620 BCE. The cult of Imhotep clearly survived into classical times: our priestly scribe Hor records in 160 BCE that he was instructed by the 'priest of the chapel of Imhotep' in the sacred city of Heliopolis.[11]

For all these reasons, there is no doubt that the minds behind the Hermetic books were Egyptian, even if they chose to express themselves in the *lingua franca* of the day. But who were they?

In the second half of the twentieth century a number of historians began to argue that the Hermetica are the 'bible' of an Egyptian mystery cult.

In recent decades a new theory of the Egyptian origins of the Hermetica has emerged. Rather than simply being the sacred books of a mystery cult, they were part of a concerted, and perhaps desperate, effort to *preserve* its teachings in the face of the great threat to their culture posed by the Greek hegemony. This anxiety found its ultimate expression in the *Asclepius'* Lament. By expressing their beliefs in the language and style of their cultural oppressors, there was a chance that the Egyptians' precious ideas would survive. This was all the more urgent because of the myriad streams of thought flooding together in Alexandria, threatening to submerge Egypt's own religious and philosophical traditions. Fowden points to the cities of Panopolis and (for obvious reasons) Hermopolis as centres of the Hermetic cult.[12]

net, in quo terram cum orbe lunari tanquam epicyclo contineri
diximus. Quinto loco Venus nono menſe reducitur. Sextum
deniqʒ locum Mercurius tenet, octuaginta dierum ſpacio circũ
currens. In medio uero omnium reſidet Sol. Quis enim in hoc

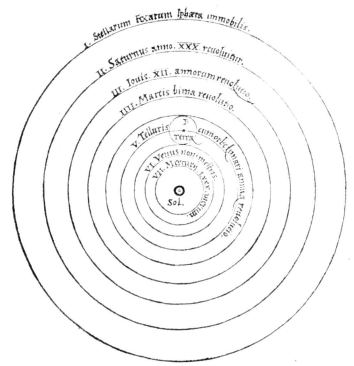

pulcherrimo templo lampadem hanc in alio uel meliori loco po
neret, quàm unde totum ſimul poſsit illuminare? Siquidem non
inepte quidam lucernam mundi, alij mentem, alij rectorem uo=
cant. Trimegiſtus uiſibilem Deum, Sophoclis Electra intuentẽ
omnia. Ita profecto tanquam in ſolio re gali Sol reſidens circum
agentem gubernat Aſtrorum familiam. Tellus quoqʒ minime
fraudatur lunari miniſterio, ſed ut Ariſtoteles de animalibus
ait, maximã Luna cũ terra cognationẽ habet. Concipit interea à
Sole terra, & impregnatur annuo partu. Inuenimus igitur ſub
hac

The famous page from Copernicus' *On the Revolutions of the
Celestial Spheres* (1543) showing his world-changing diagram
of the sun-centred solar system. Less famously, just four lines
below, he acknowledges his inspiration, the esoteric works
of 'Trismegistus' – the legendary Egyptian sage Thrice-Great
Hermes. (Bridgeman)

Detail from the lavish decoration of the Vatican's fifteenth-century Appartamento Borgia, showing Hermes Trismegistus and Moses receiving divine inspiration from the Egyptian goddess Isis – somewhat unusual for a pope's personal rooms. But this does show the extreme veneration that even the head of the Church accorded the demi-god of the Hermeticists.
(Author's collection)

The belief that Christianity could trace its origins via Hermeticism to ancient Egypt was taken to its extreme by the uncompromising Giordano Bruno, whose statue now stands on the spot in Rome where he was burned to death by the Church for heresy in 1600.
(Science Photo Library)

THOMAS CAMPANELLA —
De Larmessin sculp.

Bruno's belief that Copernicus' new model of the solar system would literally trigger a new age of spiritual and scientific enlightenment was shared by his successor Tommaso Campanella *(left)*, who in turn was a close friend and advisor to Galileo *(below)*. Considered science's great martyr because of his persecution by the Church, the evidence indicates that Galileo was motivated at least as much by the Hermetic significance of heliocentricity.
(top: Mary Evans Picture Library; bottom: Bridgeman)

Still standing tall in Rome, these ancient Egyptian obelisks re-erected in the 1650s by the remarkable Hermetic Jesuit Athanasius Kircher and the famous sculptor Bernini are replete with Hermetic symbolism. Their monument (*above left*), erected

Obelisque de la Place Navone.

outside the church where Bruno and Galileo were condemned, draws its symbolism from the extraordinary esoteric work *Hypnoerotomachia Poliphili (above right)*.

(both images: Bridgeman)

Similarly, the Fountain of the Four Rivers in Rome's Piazza Navona, encodes Hermetic secrets, as does Kircher's book on the subject, *Obeliscus Pamphilius (its frontispiece, opposite)*.

OBELISCVS
PAMPHILIVS

Io. Ang. Caninius Rom. inuentor et del.
C. Bloemaert sculp.

(opposite bottom: Herzog August Bibliothek Wolfenbüttel: A: 13.1 Eth. 2°;
above: Herzog August Bibliothek Wolfenbüttel: A: 66.1 Quod. 2°)

Sir Isaac Newton (1642–1727), whose work on gravity and the laws of motion was set out in the *Principia mathematica*, effectively created the modern technological world. Although the most famous scientist in history, he was utterly dedicated to the magical Hermetic tradition, whose principles actually drove his masterpiece. (Science Photo Library)

Although long doubted, recent research has shown that the Hermetic texts do indeed have their roots in ancient Egypt. Not only is Hermes Trismegistus the Egyptian god Thoth *(above left)* but the books encapsulate the wisdom of Heliopolis, magical city of the sun. This was the religion that inspired the building of the pyramids of Giza *(above right)* and the world's oldest religious writings, the Pyramid Texts *(below)*.
(top left: Bridgeman; other images: author's collection)

(*right*) The early universe as captured by the Hubble Space Telescope. The understanding emerging from modern physics is rapidly converging with the ancient Hermetic vision of the cosmos as the mind of God – and, to the Hermeticists, therefore also the mind of humankind. The notion of the 'participatory universe' developed by the eminent American physicist John Archibald Wheeler (*below*) intimately links human consciousness to the creation and growth of the universe, as depicted in his famous 'U and eye' diagram (*right*). In the act of observing the universe, intelligent life is in some way actually creating it.
(top: NASA; middle: Keith Prince; bottom: Science Photo Library)

By the time the Romans took control of Egypt, the Greeks had been in charge for generations, so their culture was already entrenched at the top echelons of society. But the conquerors and conquered largely kept their distance. The Greeks regarded their culture as more advanced, while the native Egyptians saw their civilization as older and wiser. The religious and cultural resistance to the Greeks was embedded in the city of Memphis, the ancient capital, whose western plateau was the 'Libyan mountain' mentioned in the Lament, site of a great necropolis that includes Saqqara where it was believed Imhotep himself was buried.[13]

The native Egyptian cults survived until Christianity became the dominant religious force in the Empire. Although the Emperor Constantine famously gave it imperial recognition, it was only in 380 CE, more than fifty years later, that Theodosius I declared it to be the one legitimate religion of the Empire. Eight years later he ordered the pagan temples of Egypt and elsewhere in the Middle East to be closed down forever, a task enthusiastically overseen by Theophilus, the Christian patriarch of Alexandria.

The Hellenic period also produced some hybrid cults that adapted traditional Egyptian worship to be more Greek-friendly. A major example of this is the cult of Serapis, a new version of Osiris worship – 'Serapis' being a conflation of the Egyptian Asar (Osiris in Greek) and Apis, the bull-god assimilated to him. The origins of the cult are controversial: was it, as long believed, a complete invention of Greek times or, as evidence now suggests, a pre-existing religion that was merely adapted for the purpose? Wherever it came from, the early Ptolemaic rulers adopted it as the official cult that could be practised jointly by their Greek and Egyptian subjects. The main temple, the Serapeum, was located in the new coastal city of Alexandria, which was founded by the Greeks in honour of Alexander the Great. However, Theophilus' over-zealous thugs destroyed it during the anti-pagan pogrom of 392 CE.

For several reasons, the cult of Serapis is a good candidate for the school that produced the Hermetica. The writers would have been associated with a temple, since in Egypt not only did learning and religion go together, but so did temples and libraries. The 'daughter' library of Alexandria's celebrated library was housed in the Serapeum – revealing the extent to which the cult valued the preservation of knowledge. The appearance in the Hermetica of Agathodaimon, the patron god of Alexandria associated with Serapis, also suggests that there was a connection with the same cult.

There certainly were some Egyptian priests who made an effort to explain their religion to the Greeks, probably in an attempt to preserve it. The major example of this is the Heliopolitan priest Manetho who, in the early third century BCE – under the first Ptolemaic rulers – wrote a history of his people, the *Aegyptica*, which is still a particularly useful sourcebook on the reigns of the various dynasties (a term he invented). Manetho is a Greek rendering of his name, but the syllable 'tho' probably derives from Thoth (perhaps 'Beloved of Thoth'), perfect for a scribe and historian of the great wisdom-centre of Heliopolis. Manetho was apparently also a key figure in establishing and promoting the Serapis religion, such was his desperation to make his people's traditions understandable to Greeks.[14] If they knew them they might like them, and if they liked them enough, they just might want to conserve them.

Manetho's agenda was the same as the one Garth Fowden ascribes to the authors of the Hermetica, which at the very least shows that some Egyptian priests were pro-actively trying to preserve their traditions.

In an ironic twist, a text ascribed to Manetho may – if

genuine – contain the earliest reference to Hermes Trismegistus. This is found in a dedication to the ruler Ptolemy Philadelphus at the beginning of the astronomical *Book of Sothis*, which is attributed to Manetho. Although this would make a particularly satisfying connection between the great Egyptian chronicler and the Hermetica, unfortunately most historians regard the book as a later work and the dedication a forgery because, following Casaubon, the term 'Trismegistus' is thought to have been invented in the early centuries CE, and therefore Manetho could never have used it.

If an attempt to preserve the Egyptian traditions was what underpinned the Hermetica, then clearly its religious and cosmological ideas would hardly have been new. They must have been the key philosophy in a belief system that predated the Greek conquest, perhaps by many centuries. Evidence for this is found in the works of the Neoplatonic philosopher Iamblichus of Syria (*c.*245–*c.*325 CE) who studied in Athens before founding his own academy in Antioch. His major work *On the Egyptian Mysteries* (*De mysteriis Aegyptiorum*) opens with the words:

> Hermes, the god who presides over rational discourse, has long been considered, quite rightly, to be the common patron of all priests: he who presides over true knowledge about the gods is one and the same always and everywhere. It is to him that our ancestors in particular dedicated the fruits of their wisdom, attributing all their own writings to Hermes.[15]

So Iamblichus understood that not only did the priests attribute their books on the nature of the gods and universe to Hermes, but also that this was already a venerable tradition that dated from the era of the 'ancestors'. As Iamblichus lived very close in time to the writing of the

Hermetica – which he frequently references – he is unlikely to have been fooled by an unashamed recent fabrication.

The Iamblichus connection is, to us, particularly satisfying. Modern academia labels him a Neoplatonist, but the opening of his masterwork, with words of praise for Hermes, suggests that his philosophy was in some way related to Hermeticism. He also 'made use of Hermetism in formulating his own widely influential doctrine'.[16] But the relationship has even more to reveal about the antiquity of the Hermetic cosmology.

'THE DIVINE IN THE ALL'

Neoplatonism was another product of Greek- and Roman-dominated Egypt. As with Hermeticism, the pro-classics bias meant that the fact that Neoplatonism developed in Egypt was considered irrelevant. Instead, scholars assumed that it was actually built on Greek ideas. However, recent studies have shown that Neoplatonism, too, owed far more to Egyptian traditions than previously acknowledged.

The 'neo' or 'new' part of the entirely modern term Neoplatonism refers to the re-establishment of Plato's Academy in Athens by fourth-generation followers of the Egyptian philosopher Plotinus. The original academy provided a meeting-place for philosophers in a grove sacred to the goddess of wisdom Athena, a mile outside Athens, for 300 years until it was destroyed by the Romans when they besieged the city in 86 BCE. With their usual disregard for local sensibilities, they cut down the sacred trees to make siege engines.

Five hundred years later, in the early years of the fifth century CE, a group of philosophers led by Iamblichus' pupil Plutarch of Athens, who considered themselves Plato's intellectual heirs, founded a new Academy in Athens. This became a renowned centre for learning in its own right, but being a pagan school it was closed down on the orders of the Emperor Justinian in 529.

The revived Platonic academy was particularly interested in exploring and developing some of the metaphysical aspects of Plato's teaching. Following his own mentor Socrates, Plato distinguished between the material and spiritual worlds, arguing that the material world, which is knowable through our five senses, is basically an illusion. Everything belonging to the material sphere is a kind of shadow of a perfect, ideal form – an archetype – that exists in the spiritual realm. Plato thought that it is possible for human beings, through intellectual effort, to transcend their perception of this world and gain experience of the spiritual realm, thus becoming enlightened.

In *Timaeus*, written around 360 BCE, Plato also introduced the concept of the Demiurge, the creator-god of the material universe. Just as everything in the physical world is a reflection of its eternal ideal, so the Demiurge is a lesser version of the one great God who created everything – including the Demiurge himself, whose power is necessarily constrained by the limitations of matter.

It was these aspects that the revived Academy was most interested in, focusing especially on the process of enlightenment through direct experience of the normally hidden spiritual realm. Rather than purely intellectual exercises of the kind advocated by Plato, the new wave of philosophers attempted to develop rituals and other practices ('theurgy') to enable the human soul to find its way back to its divine source during life rather than after death, aiming at 'the purification of the soul from the barnacles of matter'.[17] This objective was encapsulated in the last words of Plotinus (*c*.205–270), who is regarded as the founder of Neoplatonism: 'Strive to give back the Divine in yourselves to the Divine in the All'.[18]

Plotinus is an odd character. All he allowed to be known about himself comes from his pupil and biographer, Porphyry of Tyre, who also organized his fifty-four treatises into six collections of nine, hence the 'Enneads', or 'group of nine'. Plotinus was born and lived in Egypt until he moved to Rome at the age of about forty; he never revealed even to those closest to him anything about his origins or parentage. He no doubt picked up the habit of secrecy from his own master Ammonius Saccas – Ammonius the Porter – who tutored him for eleven years in Alexandria. It was from Ammonius that Plotinus learned his 'Neoplatonism'.

Unsurprisingly, Ammonius is another oddity. Described by one modern historian as 'the most shadowy figure in the chronicles of Hellenic philosophy',[19] virtually nothing is known about his life except his name, which was derived from the god Ammon, strongly suggesting he was a native Egyptian. Ammonius was known as *theodidaktus*, 'God-taught', which might be another way of saying he was divinely inspired. In any case, it suggests that his knowledge owed no debt to any formal school of philosophy recognized by the Greeks.

Ammonius Saccas set down nothing in writing, as was the custom for Egyptian priests, and placed his students under a vow of secrecy not to publish his lectures. But he had two disciples who left their mark on history, Plotinus and the Christian philosopher and theologian, Origen. It was through the latter – who apparently broke his vow – that Neoplatonic ideas were imported into Christian theology.

Mystery man he may have been, but it is still clear that Plotinus' philosophy owed more to an indigenous Egyptian source than it did to Plato. But this background cut little ice with historians of philosophy, again because of the scholarly bias in favour of the classical world. The logic behind the label 'Neoplatonist' is that Ammonius Saccas taught Plotinus, who taught Porphyry, who taught Iamblichus, who taught Plutarch of Athens, who re-established the Platonic Academy – so they all must have been Platonists, mustn't they? And in any case they were all Greek(ish), or at least very Hellenized and admirers of Plato, who was definitely Greek.

However, in the early decades of the twentieth century the avoidance of Egyptian tradition was becoming embarrassing. Even the most conservative academics had to acknowledge that large parts of Plotinus' work had no parallels with earlier Greek thought and seemed to derive from some other tradition entirely. While it is true that his writing does contain many references to Greek philosophical concepts, particularly Plato's, when these are removed, his basic principles and reasoning hold their own internal logic.[20] In other words, he may have used the Greek concepts to bolster his philosophy, but didn't *derive* it from them.

French historian of philosophy Émile Bréhier was one of the first to suggest in the early 1920s that Plotinus wasn't purely inspired by Greece – causing a huge furore among the ranks of venerable beards. At the end of his life in the 1950s, in an introduction to an English translation of his original papers, Bréhier cheekily dropped in a quote from *Asclepius*, hinting that he recognized a relationship between Plotinus and the Hermetica:

> After Alexander the Greeks, without doubt, did 'Hellenize' the Orient; but, inversely, Egypt, 'the land where gods are invented', stamped its powerful imprint not only upon the customs but upon the ideas of the Greeks, in spite of the efforts of the rulers of Egypt to keep the Egyptians in a subordinate state.[21]

But even once the non-Greek origin of much of Plotinus' work was recognized, historians still tried to ascribe his source to Iran or India – anywhere but Egypt. One might have thought that Plotinus being an Egyptian taught by an

Egyptian in Egypt might have been a clue to the source of the non-Platonic parts of his philosophy.

More recently a dose of objectivity, not to say common sense, has been injected into this unnecessarily complicated subject. Karl W. Luckert, the German-born American professor of the history of religion at Southwest Minnesota State University, has strongly and persuasively argued that Plotinus' philosophy should not be called Neoplatonic at all, but 'neo-Egyptian'.[22]

Luckert shows that Plotinus derived his ideas from traditional Egyptian spirituality. For example, he taught that the human soul comprises of both the high soul and the low soul. Not only is there nothing that corresponds to this idea in the Greek religion, but Plotinus' description matches exactly the well-known Egyptian concept of the *ka* and *ba*. The *ka* is a kind of astral double, the life force that is born with the individual and which returns to the gods at death; the *ba* is the spiritual part of the personality, the *ka*'s manifestation in the physical world. The latter is more like the traditional Western concept of the spirit body, but in the Egyptian system both make up the human soul.

Luckert goes on to show that many of Plotinus' concepts – the nature of the godhead, the human soul and its relationship to the divine – are directly lifted from Ancient Egypt. While Plotinus did use Platonic ideas, he only did so to present Egyptian traditions in a way that was familiar to his scholarly readers:

Plotinus has given us Egyptian religion, theology in the linguistic garb of Hellenic philosophy. His philosophical and Greek linguistic cover and his scarce links with Platonic philosophy sufficed to hold the attention of a few Greek students of philosophy.[23]

Further evidence of the Egyptian origins of Neoplatonism

can be seen in the career of the philosopher Antoninus, who died shortly before the suppression of the pagan cults in the 390s. Again, very little is known about him – the only source is a summary of his life written by Eunapius, an Athenian physician and philosopher, in a work dating from about a century later.

Like Ammonius Saccas and Plotinus, Antoninus was secretive and evasive about the religious element of his beliefs. Eunapius tells us that after teaching at the Serapeum in Alexandria, Antoninus went to the coastal town of Canopus in order to devote himself to its 'secret rites'. Eunapius also says that because of the growing imperial hostility to the religion, while Antonius was in Alexandria he would only ever answer people's questions using Plato's philosophy, and would flatly refuse to discuss the divine or theurgy. This is enough to label him a Neoplatonist as far as history is concerned. But clearly Antoninus was something else, something Egyptian and secret – something that *was not incompatible* with Plato, but equally not necessarily actually Plato. As Eunapius writes of Antoninus:

> Though he himself still appeared to be human and he associated with human beings, he foretold to all his followers that after his death the temple would cease to be, and even the great and holy temples of Serapis would pass into formless darkness and be transformed, and that a fabulous and unseemly gloom would hold sway over the fairest things on earth.[24]

Eunapius tells us that Antoninus' followers understood this as an oracle, which came to pass very shortly after he died in the persecutions ordered by Emperor Theodosius. His prediction echoes the Lament, although Antoninus could not have been its author as it was being quoted by Christian writers from the start of the century. However, he could

well have used the Lament as the basis for his own pre-
diction. At the very least this shows that the 'Neoplatonist'
Antoninus shared the attitudes and anxieties of the
Hermetic writer of *Asclepius*.

As with Hermeticism, the Neoplatonic trail leads to
native Egyptian traditions connected with the Serapis cult.
In fact, Neoplatonism and Hermeticism were natural
bedfellows – they are simply two sides of the same coin.

However, the Serapis cult itself was a relatively new
innovation, created or adapted for the all-conquering
Hellenic world, just as Alexandria was a new city built by
the Greeks. Therefore any traditions transmitted via the cult
to the Hermetica must have originated with some other
cult from some other place. But what was it and where did
it come from?

Karl Luckert traces the origins of the wisdom tradition of
which Plotinus was heir not just a few centuries back into
Egypt's past, but all the way to its beginning. And, if
Luckert is right, given that Neoplatonism is the twin of
Hermeticism, then wherever one is found, inevitably the
other will be also.

THE SACRED CITY

After comparing Neoplatonic spirituality with the traditional Egyptian religious schools, Luckert identified its
origin as the theology of the major cult centre of Heliopolis.
This discovery leads us to another: that strangely evocative
but mysterious city also holds the key to the wisdom of the
Hermetica. Even the name 'Heliopolis' is tantalizing, being
Greek for 'City of the Sun', which is probably why the term
appealed so much to Renaissance Hermeticists such as
Tommaso Campanella.

The golden city was the centre of the cult of the sun god
Ra, or Re (associated with the Greek Helios). Even during
the era of Greek domination the city still hosted a great

annual ceremony in his honour. Sadly, this sacred place is now submerged beneath a largely industrial suburb in the north of Cairo (although confusingly not the district called Heliopolis, which is in quite another part of the city), where a three-and-a-half-thousand-year-old temple was dis-covered beneath the market in 2006. The ancient Egyptians called it Iunu, which means 'pillars', a reference to the many obelisks that poked their phallic fingers at the sky, only one of which now remains – and it appears under the name of On in the Old Testament. The matching pair of red granite obelisks in New York and London (the anachronistically named Cleopatra's Needle) also came from Heliopolis. It was the most renowned centre for the preservation of Egypt's wisdom tradition, and the most ancient. Its reputa-tion is attested by the fifth-century BCE Greek historian Herodotus, who visited the city 'where the most learned of the Egyptians are said to be found', and met its priests.[25]

In fact, Egypt has always exerted a powerful hold over the imagination, certainly because of its literal magic and mystery but also possibly simply because its civilization endured virtually unchanged for a staggering length of time, from the unification of the two kingdoms of the Nile, Upper and Lower Egypt, around 3100 BCE. Within a mere five hundred years it had reached the level of architectural and engineering genius embodied in the great pyramids of Giza, Saqqara and Dahshur. To put these staggering achievements in perspective, this was about four thousand years before the building of the great Gothic cathedrals in European cities such as York in the north of England and the French capital.

Yet the essential aspects of ancient Egyptian civilization – its political and social structure, culture, symbolism and religion – remained more or less the same for over two millennia. Although throughout its remarkably long history there were periods of foreign occupation, the culture always rose again with its traditions basically intact. Indeed, when the Greeks took over in the fourth century BCE Egyptian culture was still recognisably the same as it always was. Even then, it continued beneath a veneer of Hellenization for another seven centuries until finally being wiped out by the Christians.

The earliest known religion of Egypt, the one that inspired the pyramid builders and other Egyptian geniuses, was that of Heliopolis. Over the course of the civilization's lengthy history other cults and religions came to the fore at different times. A particularly strong challenge was posed in the fifteenth century BCE by the religion centred in Memphis that featured Ptah as its creator-god. But the Heliopolitan theology influenced all those that came after it – it was 'the dominant strain of thought by which subsequent Egyptian religious notions and rites were oriented'.[26] The Ptah cult, for example, assimilated the Heliopolitan tradition rather than attempting to displace it. Although, like every other system of belief, it went through changes and evolved, the essential ideas remained unchanged over time. Only one attempt was made to eradicate the Heliopolitan religion completely. In the fourteenth century BC the 'heretic pharaoh' Akhenaten tried to replace it with the cult of his single solar god Aten, who was in many respects his divine alter ego.

The great genius Imhotep was a priest of Heliopolis – his cult was still practised in Heliopolis in the second century BCE, an astonishing two and a half millennia after his earthly life – and so was Manetho some 2,300 years later. Both show that Heliopolis was both a religious centre and a place of learning and science. Both priests also have a connection, albeit indirect, with the Hermetica. The prominence in the texts of Asclepius, a thinly-disguised Imhotep, suggests an association with Heliopolis. And given that Manetho was a priest of Heliopolis and instrumental in founding the Serapis cult – with which the Hermetic works seem connected – one can clearly see a bridge between the two cults.

Although the Heliopolitan religion was complex and sophisticated, nowhere in its lengthy history did its priests record its basic theology and practices. This was not the Egyptian way. Apart from an apparently engrained instinct to maintain secrecy – possibly because only the worthy could be initiated into the mysteries – priests preferred to express their religion through ritual and symbols and myths, the best-known of which is the story of Isis and Osiris.

The supreme expression of the Heliopolitan religion is the Pyramid Texts, which were inscribed in hieroglyphs on the walls of the burial chambers of the pyramids of seven pharaohs and their queens, between about 2500 and 2200 BCE. The inscriptions consist of a series of several hundred incantations relating to the afterlife journey of the deceased. However, even though the inscriptions are the most ancient religious writings known from anywhere in the world, unquestionably they derive from even older texts, dating from the very beginning of the Egyptian civilization.[27]

Yet although even the Pyramid Texts fail to set out the beliefs of Heliopolis systematically, why should we expect them to? After all, the people who mattered – the priests and worshippers – were already familiar with their own religion. The Texts do, however, allow the core theology and cosmology behind them to be reconstructed. The most successful attempt is found in Karl Luckert's *Egyptian Light and Hebrew Fire* (1991), which isolates two related aspects: the overall understanding of the origins and nature of the cosmos, and its relationship to human beings.[28]

The religion centres on nine major gods, who became known later in Greek as the Great Ennead (a group of nine). The Nine – along with a multitude of lesser gods – were all considered manifestations of the one great creator-god, Atum. The other god-forms are convenient symbols for his different aspects.

As might be expected of such a red-blooded ancient people, the Heliopolitan creation myth – or metaphor – is highly charged sexually. Luckert describes it as 'porno-graphic theography'.[29] In the original version of the creation myth Atum masturbated, his explosive climax shooting the universe out into space. Later this description was modified in order to appease the easily offended, having him spit or merely cry out – an image of a god creating the world through his divine word that was borrowed by both Old and New Testaments.

At first the image seems crudely schoolboyish, but it actually possesses rather more sophistication than meets the eye. For a start, it is a pretty good analogy for the ultimate act of creation and certainly conjures up an irresistibly lasting image, unlike those contained in dry-as-dust astronomical textbooks. And like many belief systems, the Heliopolitan religion saw everything in creation in terms of a yin yang-like complementary polarity – positive-negative, light-dark and so on – which is most often experienced at a human level as the relationship between male and female. In the original myth of Atum, his phallus is the male principle and his hand the female, and the first things they make are the goddess Shu and god Tefnut, their embodiments.

Some authorities, such as German specialist on myth and symbols Manfred Lurker, prefer the description that Atum 'copulates with himself using his hand'.[30] Although to the uninitiated this may sound like much of a muchness, the essential difference is that masturbation doesn't normally generate life. The point is that Atum contains within himself both male and female. And the metaphor of ejaculation encapsulates the notion of the universe as having a single point of origin in space and time, from which everything explodes outwards – a very modern image for the beginning of the universe. One Egyptologist even uses the word 'singularity' to describe this concept.[31]

So the universe expands outwards from Atum's own big bang, becoming not just larger but ever more complex and multidimensional, each level being represented by new pairings of deities. The first new gods or forces that come into being are the female Shu (corresponding to life) and the male Tefnut (order), who are locked in 'perpetual sexual union'.[32] This produces their more visible manifestations – in sharper focus, as Luckert puts it – the Earth-god Geb and Sky-goddess Nut. They in turn give birth to two pairs of twin gods, Osiris and Isis, and Set and Nephthys. All together they make up the Great Ennead, arranged in four levels of being, beginning with Atum. As Luckert notes:

> The entire theological system can be visualized as a flow of creative vitality, emanating outward from the godhead, thinning out as it flows farther from its source. Along its outer periphery this plethora of divine emanation becomes fragmented into what begins to appear as the light and shadow realm of our material world. It becomes visible.[33]

This is by no means the end of the process, however, as the system is repeated on a lesser octave, beginning with the child of Isis and Osiris, the falcon-god Horus, who occupies a transitional space between the Great Ennead and the Lesser Ennead, the nine gods of this world (which includes Thoth). Horus' relationship with the material universe is the same as Atum's with all creation, making him therefore the god of the material world (besides being 'a son of God and savior of humankind'),[34] the equivalent of

Plato's Demiurge and the Hermetic second god, while (like his father Osiris, who died and rose again), also bearing some comparison with Jesus.

As we have seen, according to Heliopolitan beliefs, the material universe that we perceive through our senses is only part – the outer edge, as it were – of an unimaginably vaster creation, much of which is hidden from us. Again, there is an obvious parallel with Plato's later concept of spiritual and physical worlds, which is why the last heirs of Heliopolis, the Neoplatonists, found his philosophy so suitable for their purposes.

In his book, Luckert makes a detailed comparison of the Heliopolitan theology in the Pyramid Texts and the principles of the Neoplatonists, particularly Plotinus, and finds the two are identical. Given the overwhelming evidence that Plotinus derived his teachings from secretly-taught Egyptian traditions, there seems to be no other explanation than that the Heliopolitan system was transmitted down the ages until it reached Ammonius Saccas and other Egyptian sages.

Given the close relationship between the inaccurately named Neoplatonism and Hermeticism, the Heliopolitan system must therefore also underpin the latter. And an examination of the basic ideas of Hermeticism bear this theory out. The language may be different, but the fundamental principles are the same.

VISIBLE AND INVISIBLE GODS

In the world accessible to human perception, the sun god Ra was deemed to take a role analogous to Atum's in relation to the universe as a whole, and was even known by the composite name Atum-Ra. (For the same reason Ra was also equated with Horus, as Ra-Horakhty.) It is not known when or how this conflation of the two great gods took place, only that they were being associated in the very earliest days of the Egyptian civilization. Atum was a hidden, invisible god; Ra, the golden royal sun, is his visible manifestation. This reveals a connection with the words of *Asclepius* quoted by Copernicus, namely that the sun is a 'visible god', which of course implies the existence of one that is invisible. This has an important implication: if Atum is the centre of creation, then the sun is the centre of the cosmos that humans can perceive.

There is something else that Atum conceals but implies by his very presence. The gods of both Enneads, besides representing deities concerned with specific aspects of nature and human endeavour, are all really aspects of Atum. Moreover, not only is everything created by Atum, but it is also created *from* him, which makes his creative energies and forces present in everything. Effectively Atum *is* the universe, as well as possessing a part, or energy, which lies outside and transcends it.

Human beings, too, contain Atum's 'divine spark' within themselves, making them just as god-like as the likes of Horus and Thoth. The only difference is that humans are locked into the world of matter in a way that the gods are not. This echoes the origin of another vital Hermetic principle: that human beings are potentially gods, and some even manage to actualize that potential. This was also, as we have seen, a central tenet of Hermeticism's philosophical partner, Neoplatonism, which focused on the journey of the soul to the divine in preparation for enlightenment, drawing on another crucial aspect of the Heliopolitan theology.

But there is something else that the myth of Atum has to tell us, something extraordinary. The creative flow from the god to the material universe is not just a one-way phenomenon. Just as it 'exhales' from Atum, it 'inhales' the life force of individuals, which then travels back to its source. Horus, therefore, also represents what Karl Luckert calls the 'turnaround realm', the point at which the life force can begin the journey back towards Atum. We might need Atum – but he also needs us.

The Pyramid Texts are concerned with those rituals that ensure the return of the King to Atum after his death, projecting his soul into the stars. It is commonly assumed that this stellar existence and the ability to commune with the creator is a prerogative of the King alone, becoming his only after death. However, neither is necessarily the case. The Pyramid Texts are specifically concerned with the King because they happen to be in royal tombs, but nowhere do they say that this afterlife is reserved for him alone. Indeed, the logic of the Heliopolitan theology, in which every individual is a manifestation of Atum, suggests that it happens to everyone.

The 'return journey' described in the Pyramid Texts refers to the afterlife simply because, again, they are in a tomb. But as with most other cultures, it was also believed that certain special individuals – priests or shamans, for example – could undertake this journey in life (usually in an altered state of consciousness), gaining insight or illumination.[35] This journey was also the aim of the Neoplatonists.

Remarkably, the cosmology of the Hermetica is, ultimately, also that of the first flourishing of the Ancient Egyptian culture. The belief of Renaissance Hermeticists such as Bruno and Newton that the Hermetic works represented the wisdom of that great civilization is absolutely vindicated. And Isaac Casaubon – whose work is still trotted out to trash the value of the Hermetica – was just plain *wrong*.

Other researchers have recognized the connection between the religion of Heliopolis and the Hermetica, as can be seen from the subtitle of Timothy Freke and Peter Gandy's 1997 translation of extracts from the Hermetica:

The Lost Wisdom of the Pharaohs and their translation throughout of 'God' as 'Atum'.

Of course, the obvious big question is where did the priests of Heliopolis get their ideas from? Did they dream them up, getting lucky with material that just happened to be scientifically accurate? Or was their belief system based on a genuine understanding of the way the universe is organized?

Sadly, a definitive answer about the origins of the Atum religion remains impossible because of a lack of relevant historical information. Some would no doubt prefer to explain the mystery as a legacy from an earlier, advanced, but lost, civilization, which would only push the question back further, not answer it. And inevitably some would conjure up the lazy if sensational notion that we can lay all these wonders at the door of ancient astronauts (a desperately non-Hermetic idea that implies human beings are just too stupid ever to have built wonders like the pyramids). But we suggest the greatest clue lies in the religion itself.

A major component of the magical worldview hardwired into humanity is that specially trained individuals can enter into a state of communion with the gods in which they are given intensely *practical* information. This idea is also the basis of the Heliopolitan 'return journey', Neoplatonic theurgy, the Hermetic gnosis and the occult art of memory. Such communion is not to be understood as bestowing enlightenment in the Eastern sense of the ultimate goal being the achievement of a purely spiritual state – or at least not exclusively – but as providing an understanding of how the universe works in very practical ways. This practice can then be used to extend human knowledge and induce enlightenment in the western sense, as in the Age of Enlightenment.

To judge the results of this communion we have to look no further than the great names who found enlightenment in the Hermetica, itself the ultimate expression of the ancient Heliopolitan system. Encouraging the belief that all things are possible means that the most ambitious dreams can actually be lived – and often for the greater good.

CHAPTER EIGHT

LAMENT FOR HERMES

Having looked beyond the historical clichés we see now that the scientific revolution, usually considered to have started with Copernicus and ended with Newton, was in fact the *Hermetic* revolution. Science emerged from the world of the occult in a very real and direct way. All the major players relied not just on the Hermetica's exhilarating image of humanity but also on its model of creation, which opened up their minds to the nature of the universe and its testable realities. Without Hermes Trismegistus we might never have reached the scientific age, or at least we might only have done so much later in our history.

Hermeticism always encouraged a scientific mindset, even if that was, from a modern perspective, inseparable from a more esoteric worldview. By the end of the seventeenth century the scientific component had been brutally torn from its arcane twin and given an independent existence, but the fact remains that modern science *emerged from* Hermeticism.

Today most people accept the simplistic notion that chemistry emerged from alchemy, and astronomy from astrology, as a new generation realized the error of the old ways and ditched 'irrational' practices in favour of what could be weighed, measured and tested. And yet, as we have seen, most of the greatest movers and shakers of both

Renaissance and even Enlightenment science did their best work because of their occult beliefs, not despite them. Their passion for the esoteric went way beyond mere eccentricity or an occasional hobby but was a source of electric inspiration. This was especially so in the case of Isaac Newton, whose world-changing theories were a direct application of Hermetic magical principles to physical phenomena.

This book grew out of our desire to set the record straight, to bring the Hermetic tradition back out of the shadows to take its rightful place centre stage in the history of western civilization and culture. The Hermetica has had a greater impact on our civilization than any other collection of texts apart from the Bible, and a greater impact on modern history than any other collection of texts *including* the Bible. Even those who dismiss all things occult and Hermetic might at least have the grace to acknowledge that without them the world would be very different, and arguably much the poorer. Science as we know it may not ever have come into existence. At the very least, the time to acknowledge our debt to the Hermeticists is long overdue.

And what achievers they were . . . The Hermetic tradition directly or indirectly inspired giants such as Copernicus, Kepler, Gilbert, Galileo, Fludd, Leibniz and Newton. As well as these big names, the tradition included figures who should be remembered as their equals but who have been relegated to history's second or third divisions: Tommaso Campanella, John Dee and, above all, Giordano Bruno. Apart from the luminaries featured in our story, the tradition inspired much else in the artistic and literary realms, including the works and ideas of Leonardo da Vinci, Botticelli and William Shakespeare – a pretty impressive list, surely, by anyone's standards.

Hermes' books played a central role in the golden age of Arabic science, which preserved the knowledge of the classical world, developed it and passed it back to Europe in the late Middle Ages. And the Hermetica was *the* mainspring of the Renaissance. Of course other ideas, attitudes and philosophies also contributed to that great flowering of the human mind and spirit – but the great tradition was what glued everything together.

Yet historians have long taught that other elements, such as the renewed interest in classical philosophy and learning, were at the core of the Renaissance. Hermeticism was grudgingly acknowledged, if at all, as a contributory factor, often hidden behind by the more familiar but off-puttingly dry 'Neoplatonism', or the slightly more interesting but vague label 'humanism'.[1] But an objective examination of the motivations behind the great names of the period shows the opposite to be the case. The Hermetic philosophy was at the core of the Renaissance: it was the other factors, such as a renewed passion for the works of the ancient Greeks, which were of secondary importance – and often a poor second at that.

Hermes' influence also continued as the Renaissance matured into the Age of Enlightenment, drawing to him as he did some of the new era's greatest intellects, including Newton and Leibniz.

Most of all, however, and with a fine flourish of irony, Hermeticism *was* the scientific revolution. This is no exaggeration. Just consider the following discoveries, which all owe an eternal debt to the Hermetica:

- The heliocentric theory
- The laws of planetary motion
- The concept of an infinite universe
- The idea of other solar systems containing habitable planets
- The theory of gravity
- The Newtonian laws of motion

- The circulation of the blood
- The Earth's magnetism
- The basic principles of information theory and the basic principles of computer science

The idea that mankind was of limitless potential and could do just about anything given the desire – the very *spirit* of science – also came from the teaching of Hermes. When the likes of Richard Dawkins declare that our achievements make him proud to be human, he is (presumably) unknowingly, speaking like an ancient occultist. *Magnum miraculum est homo*! The cosmic joke is not lost on Glenn Alexander Magee, who writes:

> It is surely one the great ironies of history that the Hermetic ideal of man as magus, achieving total knowledge and wielding Godlike power to bring the world to perfection, was the prototype of the modern scientist.[2]

So why isn't the Hermetic tradition given due credit? Why is it the case, as Piyo Rattansi notes, that 'to grant Hermeticism any prominence in the history of sixteenth- and seventeenth-century science is tantamount, apparently, to challenging the rationality of science'?[3]

A major reason for today's neglect is the well-established cultural bias that favours the classical world. Another is the lack of recognition, until recently, of the important contribution of Egypt's intellectual and philosophical traditions. However, this bias does not appear to be a cause but an *effect* of the neglect of the Hermetica. Until Isaac Casaubon's damning critique, even Hermes' enemies had accepted his works as the product of the most venerable period of the Egyptian civilization. Pouring cold water on the alleged wisdom texts, Casaubon tempted scholars in the opposite direction with his message that Egypt had nothing to teach us compared to the Greeks. Had Casaubon never put quill to paper, Egypt might well have remained a focus of scholarly respect, an equal of classical Greece and Rome. Had this been the case, twentieth-century academics such as Garth Fowden and Karl Luckert would never have faced such an uphill struggle to persuade their colleagues that all of the extraordinarily powerful and inviting subjects that we have seen thus far had Egyptian rather than Greek roots. Casaubon was wrong anyway. As believed by Hermeticists all along, the Hermetica authentically preserved and transmitted the cosmology and philosophy of Egypt's pyramid age, which we believe has much to teach us – even in the digital age.

Another reason for the engrained prejudice against Hermeticism is that the study of the texts was essentially forbidden after the tradition's ambitions for religious and social reform suffered serious reversals during the seventeenth century. This came about because of a paradoxical collusion between the forces of science and religion. The Catholic Church condemned Hermeticism as demonic, both because it employed magic and its perceived political threat. For their part, Protestant intellectuals backed off from the subject largely because Catholics had made it such a point of contention. One of the consequences of the power politics of the day was that it became expedient to be seen as an occult-denier, especially when the opposite could get you burnt at the stake. But the practical necessity to play it safe effectively sucked the lifeblood from the Hermetic tradition. Men of science were thus no longer men of God – or of the spirit – and soon it seemed that the two were mutually exclusive. Scientists not only denied the very existence of their predecessors' inspiration, but also had no choice but to denigrate its source.

We saw in the story of the origins of the Royal Society the

struggle between the Rosicrucian attitude and the new impersonal mechanistic experimental philosophy. There were good reasons for minimizing the influence of magic, even in Restoration England. A campaign to lose the esoteric gained favour in English academic circles, and this led those of an overt Rosicrucian or Hermetic bent to be branded sinister – and possibly satanic. In 1659 a work based on a hostile editing of John Dee's diaries, *A True and Faithful Relation of What Passed for Many Years Between Dr John Dee . . . and Some Spirits*, was published. Written by Méric Casaubon – Isaac's son, so keeping up the family tradition – it uncompromisingly painted Dee as a necromancer in league with the Devil. While it is true that with a dodgy clairvoyant named Edward Kelley, Dee had experimented over a number of years with communications with discarnate entities, they were allegedly angels rather than demons or spirits of the dead. But Casaubon Junior's book effectively trashed Dee's reputation for centuries and also cast suspicion on those who respected and worse, *used*, the good doctor's mathematical works. This was particu- larly unfortunate as, whatever one might think of Dee's esoteric studies, his was one of the greatest mathematical minds of all time.

The move from the Hermetic studies of the Renaissance to what we recognize today as science, the great intellectual flagship for rationalism and mechanism and all other resolutely non-magicalisms, was the result of the occult philosophy splitting into two parts: the magical view of the universe and its application to the phenomena of nature. Basically the theory was junked in favour of the practice.

It is often assumed that science emerged when thinking people began to question religion. This is not so: it was a specific reaction against Hermeticism – one that was actively encouraged by those members of the Catholic Church who backed Descartes' new method. What is perhaps odd given such a momentous schism, is that it was largely an accident of history that science diverged from the ancient and much loved philosophy that inspired it.

Hermeticism as a system of thought survived the Enlightenment. But just as it diverged from science, the philosophy itself became firmly the province of the occult underground and the world of secret initiatory societies. Study of the *Corpus Hermeticum* as anything other than a historical curiosity came to be reserved for students of the esoteric and magic.

The first Rosicrucian secret societies proper, formed in emulation of the brotherhood described in the *Fama Fraternitatis* and *Confessio Fraternitatis R.C.*, appeared in Germany in the first decades of the eighteenth century, part of the burgeoning interest in Freemasonry and Masonic-style organizations. However, despite claiming to be inspired by the Rosicrucian ideal, these societies were actually the opposite, exploiting the mystique around the original invisible society to add an elitist gloss to their own image while keeping their secrets, real or imagined, to themselves.

In Britain, these underground currents that flowed through Europe resulted in the influential esoteric society the Hermetic Order of the Golden Dawn. Founded in the 1880s, it not only attracted the usual suspects – famous occultists such as Aleister Crowley and Dion Fortune – but also the likes of Irish poet and patriot W.B. Yeats and, according to rumour, the originator of Dracula, Bram Stoker. To these and many others who dealt in the symbolic keys and the secret initiations that would open up both their psyches and their minds, Hermes was a god like no other, for to follow him was to become divine oneself. He has proven himself to be equally present in the lilt and lift of language and in the fire of the cosmos.

Hermeticism survived in other, less expected ways. For example, Romantic poets such as Percy Bysshe Shelley and

John Keats breathed Hermetic fire into their works as well as into their remarkably colourful and short lives. And the influential philosopher Georg Wilhelm Friedrich Hegel (1770–1831) – whose thinking inspired Karl Marx among others – was an unashamed Hermeticist. His writings, both published and unpublished, are packed with references to masters such as Bruno – whose brilliance was the subject of Hegel's lectures – and his library included books by Hermeticists and esotericists, including Agrippa and Paracelsus. Yet it took until 2001 for a study to acknowledge his Hermetic passion. Even then Glenn Alexander Magee's *Hegel and the Hermetic Tradition* was regarded as a radical new view.

Many might think that although it a shame that the old Hermetic influence on certain important historic names is neither properly nor widely recognized, surely the big split between magic and science turned out to be a good thing. After all, it allowed science to develop without the constraints of a metaphysical framework, leading to the explosion of discoveries and world-changing technologies such as steam trains, spinning jennys and telegraphy. Indeed, one could argue that Hermeticism was not necessary to make sense of this kind of scientific progress.

Up to the first half of the twentieth century, that argument might have worked. But since then science has shifted into a completely new phase, a considerably less certain world than that of Victorian nuts and bolts. And, we argue, Hermeticism is once again relevant, this time to the realm of quarks, M-theory and DNA.

As science itself becomes more magical, Hermeticism's time has truly come.

PART TWO

The Search for the Mind of God

CHAPTER NINE

THE DESIGNER UNIVERSE

The most fundamental element of the Hermetic worldview is, as we have seen, that the cosmos is not meaningless, inert or random, but is in its tiniest manifestation not only alive but also purposeful.

Unlike believers in the biblical version of creation, where God merely creates life and the universe on what appears to be a whim, to the Hermeticists as well as their ancient predecessors, the priests of Heliopolis, the material universe is nothing less than an emanation of God. In some majestically transcendent but also ultimately practical manner, the cosmos also represents his thought.

Obviously this isn't the way that the vast majority of modern scientists – as exemplified by Richard Dawkins and Stephen Hawking – see it. But we argue that it should be. Scientific cosmology has amassed a great deal of evidence about the nature of the universe that has seriously jolted the complacency of determined rationalists. The new data reveals a universe that is not merely the result of the blind workings of the immutable laws of physics. This universe emerges as being deliberately designed for a purpose in which intelligent life plays a crucial, if not *the* crucial, role.

The road to this point began back in the late 1970s when a paper appeared in the respected journal *Nature*, sending strong ripples through the scientific community worldwide.

This was entitled 'The Anthropic Principle and the Structure of the Physical World' and was written by British physicists Bernard Carr and Martin Rees. Based on the evidence of seven decades, the authors reflected on an unsettling pattern that was emerging from the accumulated discoveries of science: to an uncanny degree, the laws of physics seem to have been 'fine tuned' to allow the development of intelligent life.

Carr is now Professor of Mathematics and Astronomy at the University of London and, unusually for today, a member – and former president – of the Society for Psychical Research. Rees is the Astronomer Royal, Baron Lees of Ludlow, and since 2005, President of the Royal Society. The passage of time has done nothing to sway the authors of the paper from their original conclusions. Carr was still saying in 2008 that judging by the fine tuning, 'the universe is designed for intelligence'.[1] He is not alone. Leading cosmologists John D. Barrow and Frank J. Tipler similarly declared that:

> there exist a number of unlikely coincidences between numbers of enormous magnitude that are, super-ficially, completely independent; moreover, these coincidences appear essential to the existence of carbon-based observers in the Universe.[2]

Carr and Rees adopted British cosmologist Brandon Carter's term, first used in the 1960s of 'anthropic [man-centred] principle' to define the situation their paper examines. Carter mused about what the universe would be like if the laws of physics were different, and realized that for almost every variation, the universe they produced would be incapable of supporting life. But he later regretted 'anthropic', which refers only to humans; he had meant that the universe seems fine-tuned for intelligent life in general.

Of course, the notion that the universe was 'designed' for anything, let alone us, is unconscionable to the vast majority of scientists, since it contradicts the very basis of their discipline. Not only does it reintroduce the notion of a creator god but also the idea that the human species has some special relationship with Him/Her/It. As leading theoretical physicist Leonard Susskind remarked:

> This idea is of course anathema to physicists, who see the existence of themselves as an accidental property of a universe determined by mathematical principles, to be discovered by disinterested analysis.[3]

One can hardly imagine a more nihilistic worldview than that expressed by another theoretical physicist and Nobel Prize-winner, Steven Weinberg: 'The more the universe seems comprehensible, the more it also seems pointless.'[4]

Of course, Carr and Rees were emphatically not claiming that they had found scientific evidence for the existence of God. They were highlighting a question that science had largely avoided, having only been explored by a handful such as Carter, and then only tentatively. The anthropic principle merely makes the observation that life could never have arisen except under very specific conditions, and does not necessarily propose that they were put in place *in order* to produce life. The assumption behind Carr and Rees' paper was that what looks like design is really an illusion based on our human-centred perception of the cosmos: if the laws of physics were any different there would be no life to ponder this question in the first place. After all, just because we live on a habitable planet, it doesn't mean that the planet was created especially for us.

But they admitted that the odds were far too high to dismiss all the examples of apparent fine-tuning as coincidence. Some other, unknown, factor had to explain the illusion. As they concluded after surveying the many conditions that seemed so convincingly contrived:

> One day, we may have a more physical explanation for some of the relationships discussed here that now seem genuine coincidences . . . However, even if all apparently anthropic coincidences could be explained . . . it would still be remarkable that the relationships dictated by physical theory happened also to be those propitious for life.[5]

Perhaps this situation can be explained using the analogy of a lottery: if we win, we might ascribe our success to our skill in picking the numbers or believe we were somehow 'meant' to win, but in fact our triumph would be entirely due to chance. Much the same, the anthropic principle shows that the odds seem to have been stacked in life's favour, as if after scooping the jackpot we found that only our own numbers had been put into the machine.

Although the overwhelming majority of scientists believe that the rigging of the universal lottery machine can be explained purely in terms of an illusion – the 'weak anthropic principle' – there are some who ascribe to the 'strong anthropic principle', which stipulates that the universe is the way it is specifically to give rise to intelligent life. Among them is Freeman Dyson, the British-born American theoretical physicist, who wrote in 1979:

> The more I examine the universe and study the details of its architecture, the more evidence I find that the universe in some way must have known we were coming. There are striking examples in the laws of nuclear physics of numerical accidents that seem to conspire to make the universe habitable.[6]

'A MONSTROUS SEQUENCE OF ACCIDENTS'

In fact, the apparent fine-tuning of the universe involves so many factors that it is not merely the equivalent of winning the lottery once. This is scooping the jackpot week after week for several years.

The classic example of the fine-tuning is the origin of carbon, one of the most abundant elements in the universe, which is essential for the existence of life (as in the familiar phrase 'carbon-based life forms'), at least as far as we can conceive it. Like all but the very simplest elements, carbon is manufactured in the centre of stars, the only places hot enough, at several million degrees, to allow the nuclear fusion that, in a literal transmutation, builds the atoms of one element from those of others. At the beginning of the 1950s, scientists understood the principle behind the formation of carbon, but not precisely how the process worked, as there seemed to be an insurmountable obstacle. Although an atom of carbon is made from the fusion of the nuclei of three atoms of helium, this should be an extraordinarily rare event, as first *two* helium nuclei had to fuse, and the resultant structure (an atom of beryllium) is so unstable it should be impossible for it to survive long enough for a third nuclei to join the act. If carbon managed to exist at all, it should be a very rare element, whereas of course the universe is actually overflowing with it. Clearly, some kind of special condition exists that increases the chances of the three helium nuclei coming together.

In 1951 a British astronomer, the celebrated – and to some, notorious – scientific maverick Fred (later Sir Fred) Hoyle, speculated that the solution to the mystery sur-rounding carbon was that the nucleic energy is enormously amplified by an aspect of quantum theory called resonance. This would prolong the life of the beryllium and so greatly increase the chances of the third helium nuclei joining the party. From this premise, Hoyle was then able to calculate what the energy of the resonance ought to be. An American team at the California Institute of Technology (Caltech), the only place at that time where such experiments could be carried out, tested Hoyle's prediction and found he had been precisely correct. This was a watershed event in the modern history of science, marking an enormous leap in the understanding of the way all elements are created. But while the American team were honoured with a Nobel prize for the discovery, blunt Yorkshireman Hoyle was overlooked (as we will see in the next chapter), almost certainly because by the time the prizes were awarded in the mid-1980s, Hoyle had championed two controversies too far – the theory of panspermia, the idea that life came to Earth from space, and that of the 'intelligent universe'.

What really intrigued Hoyle was the precision of the energy 'spike' produced by the resonance, known as the triple-alpha process. If it were just one per cent higher or lower the reaction would fail, ultimately leaving only tiny amounts of carbon in the universe, and therefore no life. As there seemed no reason for the resonance to be so precise *except* to make the process work, Hoyle went so far as to describe it as a 'put-up job'.[7]

The significance of the triple-alpha process goes beyond the creation of carbon, since all other elements necessary for life depend on it. Stellar 'factories' work by adding nuclei to the atoms of one element to make a new, more complex, one. Just as beryllium atoms have to form before carbon can be made, so carbon atoms are needed to make oxygen, oxygen atoms to make neon, and so on. All these reactions are more straightforward than the triple-alpha process as they don't require the convenient energy spike, so the obstacle Hoyle faced isn't present. But if carbon did not exist, then neither would any of the elements above it on the periodic table. Literally everything depends on the triple-alpha process. Without it there would only be four elements in the entire universe.

Such apparent contrivances prompted Hoyle to declare in a lecture at University Church, Cambridge, in 1957:

> If this were a purely scientific problem and not one that touched on the religious problem, I do not believe that any scientist who examined the evidence would fail to draw the inference that the laws of nuclear physics have been deliberately designed with regard to the consequences they produce inside the stars. If this is so, then my apparently random quirks have become part of a deep-laid scheme. If not then we are back again at a monstrous sequence of accidents.[8]

Since Hoyle made that statement, the more science has discovered about the origins and evolution of the universe the more 'monstrous' the 'sequence of accidents' has become.

One of the first to be intrigued by Brandon Carter's anthropic principle was British cosmologist Paul Davies – that rare animal, both a highly regarded academic and a successful popular science writer. He has continued to explore the implications and mysteries of the anthropic principle, most famously in *God and the New Physics* (1983) and *The Mind of God* (1992), and most recently in *The Goldilocks Enigma* (2006) – the title referring to the conditions in the universe that are, like Goldilocks' porridge, 'just right' for life.

Davies points out that life has three main requisites: 'stable complex structures' in the universe (galaxies, stars and planets rather than clouds of gas or vast numbers of black holes); certain chemical elements (for example carbon, oxygen); and a place where the components can come together (for example the surface of a planet). Of course our universe has all of these components, but each requires such fortuitous circumstances to exist that ours is, as Davies puts it, apparently a 'designer universe'.[9]

The universe as it is today is, of course, the result of how it was in the beginning. Had conditions been different then, it would be different now – and almost certainly hostile to the development of life. According to today's thinking, the universe began 13.7 billion years ago with the 'big bang'. (Ironically the term was invented by the sceptical Hoyle, but only as a put-down. Then to compound the irony, his team found some of the best supporting evidence for it.) The big bang had to be within a certain range of size and explosive potential to produce our universe. If it had been bigger or bangier, it would have expanded too quickly for galaxies to form. If it had been smaller or less bangy, gravity would have pulled the universe back into itself well before life could have evolved.

For a time after the big bang the expanding universe was too hot to be anything other than a dense, incandescent plasma composed of subatomic particles like protons, neutrons and electrons. As it expanded further it cooled, so that – an estimated 380,000 years after the big bang – the particles could fuse to form the simplest elements, hydrogen and helium. Those two elements make up about 99 per cent of matter in the universe. But if the relative masses of protons, electrons and neutrons were only minutely different, not a single hydrogen atom could form. It seems we must boldly go well beyond the frontiers of coincidence to begin to understand the way our universe was created, and how it continues to work.

Attracted by the gravity of individual atoms, clouds of hydrogen and helium clump together, clumping faster and faster as they grow. The smaller the clumps, the hotter they become, until they are hot enough to kick-start nuclear reactions – and it is at this stage that a star is born, whose deadly beauty masks its true self, a gigantic fusion reactor. Acting as unimaginably massive factories that manufacture more complex elements from hydrogen and helium, stars then disperse these into space where they explode as supernovae. Every atom in every molecule, including those that make us up, was born in a star light years away, millions or billions of years ago, making even the tiniest newborn in some respects old beyond imagining. As the legendary American theoretical physicist Richard P. Feynman observed, 'the stars are of the same stuff as ourselves'.[10] And as Paul Davies comments:

> The life cycle of stars provide just one example of the ingenious and seemingly contrived way in which the large-scale and small-scale aspects of physics are closely intertwined to produce complex variety in nature.[11]

But contrived and intertwined by whom or what?

There are also many instances where a combination of factors has to work together to produce a bio-friendly outcome – almost as if knobs are being twiddled until the balance is exactly right. In his 1999 book *Just Six Numbers*, Bernard Carr explored six of the fundamental forces, or relationships between forces, on which the universe is built. He found that all of them are very finely balanced, and if they were just slightly smaller or larger they wouldn't produce a life-friendly universe. As he pointed out in 2008:

> Known physics does not explain these fine tunings. It seems indisputable that these relationships are required in order that life can arise, and they're really quite precise – they don't determine constants uniquely, but they do determine constants to, say,

within something like 10 per cent, and there simply is no explanation.[12]

Stephen Hawking also acknowledges this remarkable phenomenon:

> The laws of nature form a system that is extremely fine-tuned, and very little in physical law can be altered without destroying the possibility of the development of life as we know it. Were it not for a series of startling coincidences in the precise details of physical law, it seems, humans and similar life forms would never have come into being.[13]

Perhaps the most astonishing example of fine-tuning comes from the most recent to be discovered. In order to describe it, we need to start from the apparently bizarre premise that there is no such thing as 'empty' space; even the interstellar vacuum is filled with 'virtual particles' that nevertheless possess energy. This has an effect on the rest of the universe, specifically on the rate at which it is expanding. In theory at least, the 'vacuum energy' has huge potential significance in terms of the anthropic principle. Some of the particles will be positive, some negative. If the sum total of all the vacuum energy were positive, then the expansion of the universe would be accelerating, and if it were above a certain value then the universe would have expanded too fast for galaxies to form – matter would fly apart faster than gravity could pull it together. On the other hand, if the vacuum energy were negative, the life cycle of the universe – from big bang to big crunch – would be too short for life to evolve.

In practice, however, the presence of vacuum energy was not considered too important – at least until the 1990s. The rate of expansion was believed to be constant, neither accelerating nor decelerating, which meant that the vacuum energy played no part in the process. This, in turn, meant it must have a net value of zero – that is, all the energy in the vacuum was neatly balanced, the positive and negative particles cancelling each other out. Cosmologists had no idea why, but that's what the data suggested.

But in the mid-1990s this happy state of affairs suffered a major jolt, as independent research based on new, more accurate data from sources such as the Hubble Space Telescope showed that the rate of expansion is, in fact, speeding up. This means that the vacuum energy has a slight positive value, not all of which is cancelled out by the negative. It is only a tiny imbalance: calculations showed that the positive energy value is 10^{120} times (that's 119 zeroes after the decimal point before you even get to the 1) less than the total positive energy in the vacuum. In other words, the negative energy cancels out all the positive – apart from a minute portion.

Learned jaws were on the ground yet again when it was realized that if that value was just one decimal place shorter – that is, the actual positive energy was 10^{119} times less than the total (or 118 zeroes after the decimal point and before the 1) – then the universe as we know it simply couldn't exist. It would expand too quickly for galaxies, stars or planets to form. Referring to this as the 'biggest fix in the universe', Davies points out that this tiny difference – a point between the 119th and 120th decimal place – is the thinness of the knife edge on which all life is balanced.[14] In answer to the dilemma posed by this 'staggeringly precise' balancing of the vacuum energy, Leonard Susskind writes: 'This seems like an absurd accident and we have no idea why it should happen. There is no fine-tuning quite like this in the rest of physics.'[15]

However, while acknowledging that there is no viable alternative to an 'anthropic explanation',[16] Susskind does not imply the existence of a 'grand designer'. For him this phenomenon can only be explained by whatever is behind the anthropic effect as a whole, which to the conventional scientist means the *illusion* of design. For Susskind, however, as for many scientists, there is only one solution to the conundrum: the marvellous and all-encompassing notion of 'multiverse'.

INSIDE THE MULTIVERSE

According to the fans of this fashionable hypothesis, there are millions or billions, perhaps an infinite number, of universes, co-existing invisibly alongside our own, each governed by its own laws of physics. We just happen to live in one that happens to be bio-friendly. It may appear to have been custom-made for us, but as this universe is, by definition, one that will sustain our sort of life and the only one we can perceive, this is the only one we know about.

The multiverse is a concept that turns the virtually impossible into the almost inevitable. To use the lottery analogy again, if your ticket automatically entered you into several million draws simultaneously, it would hardly be surprising if your numbers came up somewhere. The same logic dictates that by positing millions upon millions of universes, the odds that at least one would boast the right conditions for life are drastically shortened.

The multiverse theory is the only alternative to design that remains within the bounds of scientific credibility and allows the anthropic conundrum to be debated without professional anxiety. Bernard Carr explains that physicists regard the multiverse hypothesis as the 'legitimization'[17] of the anthropic principle.

However, unfortunately for its many scientific fans, there are major problems with the multiverse. First, and surely the most damning, is that it is purely a theory with not a shred of solid data to back it up. There are three basic, competing models of the genesis of multiple universes that may keep physicists agog with debate and busy formulating mathematical models of how they might work, but this seems a hollow exercise as none of the models have the remotest hope of ever being proved. In fact, it is impossible to gather evidence because interaction between universes is by definition also impossible.

On the other hand, multiverse theory can be used to predict certain features of *this* universe. But as American theoretical physicist Lee Smolin, founder of the Perimeter Institute for Theoretical Physics in Ontario, Canada, notes:

> Within the standard model of elementary-particle physics, there are constants that simply don't have the values we would expect them to have if they were chosen by random distribution among a population of possibly true universes . . . In fact, I know of no successful predictions that have been made by reasoning from a multiverse with a random distribution of laws.[18]

The theory also violates another highly-prized scientific principle, Occam's razor, expressed by the great twentieth-century physicist Sir James Jeans as, 'We must not assume the existence of any entity until we are compelled to do so',[19] or, in other words, the simplest explanation is usually the best. As Paul Davies wryly comments: 'To invoke an infinity of other universes to explain one is surely carrying excess baggage to cosmic extremes.'[20]

The complete absence of evidence does not justify the extraordinary confidence with which the multiverse is promoted as a solution to the anthropic conundrum. In a 2008 radio discussion, British theoretical physicist Fay Dowker stated that 'the existence of the multiverse, if we can establish it, would eliminate the question of why the laws of nature are the way we see them'.[21] *If* we can establish it . . .

In the introduction to *Universe or Multiverse?* (2007), Carr acknowledges that the multiverse hypothesis:

> . . . is highly speculative and, from both a cosmological and a particle physics perspective, the reality of a multiverse is currently untestable. Indeed, it may always remain so, in the sense that astronomers may never be able to observe the other universes with telescopes and particle physicists may never be able to observe the extra dimensions with their accelerators.[22]

He goes on:

> For these reasons, some physicists do not regard these ideas as coming under the purview of science at all. Since our confidence in them is based on faith and aesthetic considerations (for example mathematical beauty) rather than experimental data, they regard them as having more in common with religion than science.[23]

In recent years the multiverse theory has become inextricably bound up with two others: string theory and the related M-theory. These are now locked in a symbiotic – indeed circular – relationship. To put it baldly, one is taken as proof of the other. Unfortunately, however, both the string and M-theories suffer from the same problems as the multiverse. And a growing chorus of physicists are volubly expressing doubts about their validity and whether, despite all the time, effort and often almost hysterical enthusiasm devoted to them, they are nothing more than a complete dead end. One of the most withering attacks on string theory came in 2006 from Lee Smolin in his book *The Trouble with Physics*.

String theory – often called 'superstring theory' in a rather pitiable attempt to make it sound sexier – posits that instead of being single points, subatomic particles are all manifestations of a single type of vibrating one-dimensional string-like entity that expand and contract as they gain or lose energy. As they are beyond tiny, one-trillionth of a trillionth the size of an atom, obviously no one has ever seen one. They only definitely exist within the realm of mathematical formulae.

String theory was formulated in the mid-1980s and was quickly recognized as the best hope for the physicists' dream of a theory that would unify relativity and quantum theories, the grand unified theory or theory of everything. However, it rapidly moved in the opposite direction. As it failed to explain certain things, variations were suggested to account for them, and so every attempt to fix the initial problem ended with another variant of the basic theory – adding new excrescences to the equations. As the number of variations multiplied exponentially, creating new sub-theories, each with its own problems, attempts to fix those led to more variations. And so on.

Physicists belonging to the old guard reacted with alarm. Richard Feynman declared: 'I don't like that for anything that disagrees with an experiment, they cook up an explanation – a fix-up to say, "Well, it still might be true."'[24]

The numbers involved are literally beyond imagining. Based on the currently-understood value of certain cosmological parameters, when all the different variables are taken into account, there are around 10^{500} possible versions of string theory. That's 1 followed by 500 zeroes – difficult enough to write down, let alone imagine – about six times the number of atoms calculated to exist in the observable universe. As Smolin points out:

Even if we limit ourselves to theories that agree with observation, there appear to be so many of those that some of them will almost certainly give you the outcome you want. Why not just take this situation as a *reductio ad absurdum*? That sounds better in Latin, but it's more honest in English, so let's say it: if an attempt to construct a unique theory of nature leads instead to 10^{500} theories, that approach has been reduced to absurdity.[25]

In 1995 the term 'M-theory' was coined in an attempt to bring order to the chaos. M-theory simply means the single theory that is assumed to lie behind all the variations of string theory and which, once established, will reconcile them all. Although 'M' was chosen randomly – like labelling an unknown quantity 'X' – those to whom it is the ultimate answer have happily tied themselves in knots trying to work out what it means, suggesting it might stand for 'magic', 'mystery' or 'mother'. Those who are undecided about its value suggest maybe it stands for 'maybe'. Sceptics prefer 'myth'. Despite the fact that M-theory is simply shorthand for a desperately needed, but currently non-existent solution to the complex problems posed by string theory, many physicists now solemnly make statements like 'according to M-theory . . . '

In a seminal paper in 2002, Leonard Susskind, the 'father of string theory', one of those who originally formulated it in the late 1960s, proposed a unification of the string and multiverse theories that made a virtue of the vagueness of M-theory. He was compelled in that direction by the astonishing precision of the near-cancellation of vacuum energy that we discussed earlier, which he realized could only point to an anthropic explanation. Susskind proposed that every variation of string theory was as correct as any other – each simply defines the laws of physics for a different universe. In what he termed a 'landscape' of string theories, he proposed that rather than one theory of everything, there are really lots of 'everythings', each with its own theory.

So, although the term 'M-theory' was originally invented as an umbrella term for the 10^{500} competing variations of string theory, its advocates, most prominently Susskind and Stephen Hawking, have turned it into a single theory in its own right. This 'proves' that there are 10^{500} different string theories defining the laws of physics for 10^{500} different universes, and is therefore taken as proof that the multiverse is real.

This may be an ingenious exercise in explaining one unknown by another, but that's all it is. As we have seen, the multiverse theory is after all by definition untestable, and M-theory unproven to say the least. As Jim Al-Khalili, theoretical physicist at the University of Surrey, commented:

> The connection between this multiverse idea and M-theory is . . . tentative. Advocates of M-theory . . . would have us believe that it is done and dusted. But its critics have been sharpening their knives for a few years now, arguing that M-theory is not even a proper scientific theory if it is untestable experimentally. At the moment it is just a compelling and beautiful mathematical construct . . . [26]

The situation thus becomes very much like the argument between those who insist their chosen god is bigger and better than any other, a line that so rouses Richard Dawkins' ire. To him, this is ludicrous even to begin to debate, as *no* gods exist. Yet here we have a very similar attitude. The arguments about multiverses and string theory are basically *theological* debates without a god or gods.

Clearly the multiverse explanation of why we live in a bio-friendly universe is (to put it as kindly as we can) at best speculative. As Smolin comments, because the multiverse hypothesis can't be confirmed by direct observation, it can't be used as an explanation and conversely, 'the fact that we are in a biofriendly universe cannot be used as a confirmation of a theory that there is a vast population of universes.'[27] The late John Archibald Wheeler, who took on Einstein's mantle in the 1950s, discovered black holes and is widely regarded as the greatest theoretical physicist of modern times, considered the multiverse as unscientific speculation that carried 'too much metaphysical baggage'.[28]

Paul Davies explores an ironic and amusing twist to the multiverse theory, one that takes the story into rather unsettling *Matrix*-esque territory. This invokes another sci-fi idea that is nevertheless taken seriously by many scientists, that of simulated universes. Building on the ideas of British philosopher Nick Bostrom, Davies explored the implications presented by the simulated universes concept in the 'design vs. multiverse' debate.

As Davies pointed out in an article in 2003, since multi-verse theories posit an infinite number of universes, anything anyone can think of will inevitably happen in one or more of them. Although only rarely will one universe possess the right conditions for life, there will still be masses of inhabited universes. (After all, what's a small percentage of infinity?) In some of them, civilizations will have arisen that are so technologically advanced they will have developed their own computer-simulated, *Matrix*-style universes. For all we know, we might be living in one. (But how would we ever know if there were no red pills?) After all, a civilization that can simulate one universe can simulate many. As Bostrom points out, the ability to run such simulations wouldn't remain confined to a civilization's scientists, but would eventually filter down to students, schoolchildren, artists and even hobbyists. Programmers might even create universes where the inhabitants are advanced enough to simulate their own universes. The logical outcome would be that the *majority* of universes would be artificially designed.[29]

This provocative scenario does, of course, depend on the multiverse theory being correct in the first place, and Davies is far from convinced of this. The point of his paper is that *if* one accepts the multiverse, then one also has to accept that the odds are in favour of our universe actually being simulated. So, pushed to its logical conclusion, even the multiverse theory supports the idea of design!

What surprised Davies was the enthusiasm with which proponents of the multiverse such as Lord Rees took to their idea.[30] They are much more willing to accept that our universe is designed by a computer programmer than that it was designed by a God or gods – even though the distinction is of course, essentially merely semantic. To humanity the Great Programmer(s) would *be* divine and omnipotent – so they might as well be gods.

A DESIGN FOR LIFE

Even with such prestigious opponents as Wheeler, most physicists and cosmologists accept the multiverse theory. But are so many of the best modern scientific minds simply clinging to it just because they're afraid of facing the very unwelcome implications of the anthropic principle?

The evidence underpinning the anthropic principle suggests one of two scenarios: either the cosmos was intelligently designed, specifically to produce intelligent life, or there is something about it that makes it *seem* like this is the case. The only suggestion that has been made about what that 'something' might be is the multiverse. This presents us with a straight choice between one or the other.

And if the multiverse is wrong then science itself proves that the universe is designed for life.

This choice is recognized by most leading physicists such as Stephen Hawking, who writes that the anthropic principle 'suggests either intelligent design or, if there are trillions of universes as M-theory proposes, that luck and probability are enough to make our existence feasible'.[31] In his 2010 *The Grand Design*, co-written with Leonard Mlodinow, he comes down firmly on the side of the multiverse and M-theory, which led to his well-publicized pronouncement that God did not create the universe, while acknowledging that M-theory hasn't yet been proven. Jim Al-Khalili, however, points out that this is essentially the same logic as those used by religionists. While they use fine-tuning, along with their faith, as evidence for the existence of God, Hawking and his fellow advocates of M-theory seize on it – together with the assumption that there is no God – as evidence for their own hypothesis.[32]

No less a figure than Steven Weinberg, the eminent American Nobel prize-winning theoretical physicist, when discussing the enigma of the vacuum energy, writes that if further research confirms this seemingly miraculous balancing act 'it will be reasonable to infer that our own existence plays an important part in explaining why the universe is the way it is'.[33] Susskind calls Weinberg's statement 'the unthinkable, possibly the most shocking admission that a modern scientist could make: man's place in the universe may indeed be at the centre'.[34] Of course, despite those words, Weinberg, champion of the 'pointless universe', will not agree for a moment that man is at the centre of things. He goes on:

For what it is worth, I hope that this is not the case . . .
I hope that string theory really turns out to have

enough predictive power to be able to prescribe values for all the constants of nature . . .[35]

But if string theory finally and comprehensively falls, as it shows every sign of doing, then we will be left with Weinberg's reasonable inference that the presence of intelligent life is fundamental to explaining the universe. This would mean that science itself effectively provides overwhelming evidence for the designer universe, which of course means there must be a Grand Designer.

We are often told that science is an evolving, self-correcting process, in which its laws and theories are never fixed but merely contingent, the best conclusions that can be drawn from the available data. It is also implicitly understood that future discoveries may completely overturn current thinking and lead to a revision of the theories. But when it comes to the anthropic principle this reasoning suddenly falls by the wayside.

The best available data from physics – the hard facts it has amassed, which can then be tested experimentally and empirically – points unequivocally to a universe fine-tuned for intelligent life. However, the majority of scientists argue that one day we will have better data that will show this to be an illusion. But all their supporting 'evidence' is theoretical, speculative and untestable. We can imagine what would happen in any other field of human endeavour if someone admitted they had factual evidence pointing in one direction, but then declared they can think of a hypothetical reason why the opposite, which unfortunately is impossible to test, is true.

Why should this be? Why should the normal rules of science change when it comes to the anthropic principle? The justification for making it a special case is that a designed universe violates one of the most fundamental principles on which the scientific worldview and method is based. The scientific revolution, we are told, came about when thinkers realized that physical phenomena could best be explained in terms of mechanical processes and laws that are purely a consequence of the way the universe is – without presupposing the existence of a designing and guiding intelligence.

However, as we saw in Part One, this is *not* the way the scientific revolution happened. *All* of its great figures – Copernicus, Kepler, Galileo, Newton, Leibniz – based their work on the understanding that the universe *was* intelligently created and that human intelligence plays a key part in its design and purpose. Bruno even anticipated the existence of other, more advanced extraterrestrial intelligences, which fits the strong anthropic principle even more neatly. None of them would have had any problems with the implications of the anthropic principle; they would have taken it for granted. And they certainly wouldn't have tied themselves in theoretical knots to evade the evidence staring them in the face.

Opponents of design point out that the hypothesis is just as untestable as the multiverse theory. That is not the case. The hypothesis of creation by deity or deities unknown does allow for the formulation of testable predictions. What predictions? Simply, if the universe is designed for intelligent life then the more our understanding of physics advances, the more we will uncover evidence of such design. Which is, of course, exactly what has happened. The design hypothesis passes that test.

A few scientists have at least been open to the notion of some form of design. Fred Hoyle proposed that the 'intelligent universe' (the title of his 1983 book) is a purposeful, creative entity evolving towards some specific end. Hoyle also scathingly dismissed the usual scientific response to the anthropic principle, calling it 'a modern attempt to evade all implications of purpose in the

Universe, no matter how remarkable our environment turns out to be'.[36]

The most high-profile scientific advocate of the design idea now is Paul Davies, who summed his position up in *The Mind of God* (1992):

Through my scientific work I have come to believe more and more strongly that the physical universe is put together with an ingenuity so astonishing that I cannot accept it merely as a brute fact. There must, it seems to me, be a deeper level of explanation. Whether one wishes to call that deeper level 'God' is a matter of taste and definition. Furthermore, I have come to the point of view that mind – i.e. conscious awareness of the world – is not a meaningless and accidental quirk of nature, but an absolutely fundamental facet of reality. That is not to say that *we* are the purpose for which the universe exists. Far from it. I do, however, believe that we human beings are built into the scheme of things in a very basic way.[37]

However, perhaps oddly, the theory that suffers the most from the design interpretation of the anthropic principle is the traditional idea of God as creator, because it exposes the limitations of his divine power.

The God of the Judeo-Christian religion, for example, created worlds from his will and word alone, and fashioned Adam out of clay and Eve from a rib bone. This was not metaphorical, but *literal*. After that particular tour de force tweaking the resonance of the helium nuclei or making a minor adjustment to the strength of the weak nuclear force in order to produce a man and woman millions of years later is something of an anticlimax.

This is not properly understood (or perhaps it is, but evaded) by those representatives of organized religions who use the evidence for design in support of their own doctrines. We find ourselves in the unusual position of agreeing with a pope, John Paul II, in his 1985 statement that to dismiss the scientific evidence for design in the universe as being a simple coincidence 'would be to abdicate human intelligence'.[38] But we profoundly disagree that such evidence supports the existence of the God of the Bible, and therefore of the Catholic Church.

The Cardinal Archbishop of Vienna, Christoph Schönborn, supremely missed the point when he declared in 'Finding Design in Nature', published in the *New York Times* in 2005, that by refusing to accept chance explanations for the way the universe works, the Catholic Church is 'standing in firm defence of reason' and that it will again defend human nature by proclaiming that the immanent design evident in nature is real'.[39] He is quite wrong: the evidence of design *disproves* the Catholic teachings about God, and it is disingenuous to pretend otherwise.

However, the 'designer universe' concept does support the cosmology of the Hermetic tradition, as well as the Neoplatonists' and the Heliopolitan theology that we argue lay behind them. Paul Davies notes that the kind of designer suggested by the strong anthropic principle fits the model of the Demiurge – the lesser or, in the words of the Hermetica, 'second god', whose creative power is constrained by matter – rather than the omnipotent God of Judeo-Christian tradition.[40] So in this respect at least, science supports the Hermetic tradition.

At this point the exact nature of the designer isn't the most important consideration. If we have to use a term that doesn't commit us to any specific image, we suggest Grand Universal Designer – or the good GUD almighty.

In this chapter we have only explored the *conditions* that made the universe ripe for life. If GUD exists, we should be able to see evidence of his or her hand elsewhere in nature, particularly in the emergence and development of intelligent life. On the other hand, other branches of science may utterly demolish poor old GUD by demonstrating conclusively that certain phenomena could only happen through the workings of pure chance and blind forces. But which way does it go?

CHAPTER TEN

STARDUST IS EVERYTHING

In the last chapter we saw that advances in cosmological understanding point firmly in the direction of the design interpretation of the anthropic principle, suggesting that the universe was intentionally fine-tuned – by whom or what we have no way of knowing – specifically to make it suitable for intelligent life. But this only concerns the physics, the manufacture of the elements necessary for life and the planets where it can dig in and thrive. What about the next step? How are living things actually made? And do the processes that create life support the designer universe hypothesis?

After all, if life itself turns out to be an incredible fluke, the whole idea of a designer universe would be undermined. On the other hand, if the laws of physics have been rigged to produce a universe agog for life, we would expect the rules of biochemistry to be similarly primed to ensure life develops wherever and whenever it can.

Frustratingly, however, matters are not as cut and dried as they are with the physics, since there are enormous gaps in the available data. Charles Darwin wrote to his great friend, the botanist Joseph Dalton Hooker, in 1863, four years after the publication of *On the Origin of Species*, saying: 'It is mere rubbish, thinking at present of the origin of life; one might as well think of the origin of matter.'[1] Although

239

150 years later we know considerably more about the origin of matter itself, our information on the origin of life is still largely 'rubbish'. Darwin's foremost modern apostle, Richard Dawkins, writes in *The Greatest Show on Earth: The Evidence for Evolution* (2009) that 'we have no evidence bearing upon the momentous event that was the start of evolution on this planet'.[2] '*No* evidence . . .' None whatsoever.

Since Darwin took the discussion of the evolution of life to a new level in the mid-nineteenth century, biologists' growing understanding of the conditions necessary for complex life forms could be extrapolated in two diametrically opposite directions. Some still consider that the chain of events that led to life on Earth was so dependent on chance that organic life must be an extremely rare phenomenon, cosmically speaking. Some even argue that the odds are so stacked against the development of life that Earth may be unique in the universe. Yes, they claim, we are alone – get used to it. On the other hand, some believe the processes that produce life unfold according to rigid laws. What happened here will happen anywhere given approximately the same conditions. And given the vastness of the universe, even if those conditions were rarer than multiverses with life, there will still be millions of suitable locations for it to exist.

Once upon a time most biologists believed that life was an exceedingly rare phenomenon at best. But new discoveries in the last two or three decades prompted specialists to see it as a common, even inevitable, feature of the universe. A phrase that is often bandied around is that life is a 'cosmic imperative': the ordering of the universe means that wherever conditions are such that life *can* evolve, it *will*, just as weeds will seize on the tiniest nooks and crannies to grow and thrive. Life just can't stop itself.

One of the foremost exponents of this school is Christian

de Duve, the Belgian biochemist and cytologist who won a Nobel Prize in 1974 for his work on cells. In 1995 he published *Vital Dust: Life as a Cosmic Imperative*, a detailed survey of the origin and development of life on Earth, from the first organic molecules to human beings. He writes:

> . . . life is the product of deterministic forces. Life was bound to arise under the prevailing conditions, and it will arise similarly wherever and whenever the same conditions obtain. There is hardly any room for 'lucky accidents' in the gradual, multistep process whereby life originated.[3]

It may be early days yet, and the evidence may be nowhere near as conclusive as that for the fine-tuning that led to the formulation of the anthropic principle, but the very fact that the study of the origins of life, or abiogenesis, is moving in this direction is implicitly designer-universe friendly. This also fits in with the Hermetic principle that the universe is teeming with life – or at least the potential for life. Giordano Bruno took this line of thinking to its logical conclusion, arguing for the existence of other inhabited worlds.

The modern trend towards seeing life as a cosmic inevitability arose largely from the growing recognition that the universe is brimming with the building blocks of life – not just on planets but even in deepest space.

ALIEN SEEDS

The spring of 1953 was a big time for abiogenesis: two seminal scientific papers appeared within just three weeks, fuelling great excitement in the subject. The first was published in the 23 April edition of the British scientific journal *Nature*, by James D. Watson (a somewhat maverick American biologist) and Francis Crick (British physicist-turned-biologist), announcing their discovery of DNA's double helix. Then on 15 May the American *Science* carried a paper by Stanley L. Miller on his and Harold Urey's re-creation at the University of Chicago of some of the fundamental chemical building blocks of life – most significantly certain amino acids – under simulated 'primitive Earth' conditions.

At the time, it was Miller who made the bigger splash. Watson and Crick's paper was about what was then considered a very uninteresting nucleic acid, only hinting cautiously, in its very last sentence, that it might actually be the long-sought medium of genetic inheritance: 'It has not escaped our notice that the specific pairing we have postulated immediately suggests a copying mechanism for the genetic material.'[4] But despite their laid-back comment, the discovery of DNA made the scientific landscape richer, more colourful and intoxicatingly alive with promise.

Miller's paper, on the other hand, offered much more hope for unlocking the origins of life. It seemed to confirm the prevailing theory that it began in the Earth's 'primordial soup' of biochemicals. The implication was that further research would reveal how the more complicated parts of the system came into being through similar processes – all of them essentially blind.

As we now know, by unravelling the genetic mystery, Watson and Crick's discovery has had by far the greater impact, not just on science, but on our daily lives – witness, for example, the DNA 'fingerprinting' used to catch criminals. Urey and Miller haven't fared nearly so well, partly because although their experiments showed amino acids and certain other biogenic chemicals could be produced easily in the lab, taking it further and putting the building blocks together in any more complex way remained out of reach. Since 1953 it has also been discovered that creating, for example, amino acids doesn't require terrestrial conditions at all. Many of the building blocks of life have been found literally floating around in space.

For a long time it was assumed that however life on Earth originated it happened *on* Earth. Even over a century ago this was not without its challengers, however. Great names of the Victorian age such as German physicist Hermann von Helmholtz and British physicist and engineer Lord Kelvin advocated that the seeds of life could be carried between planets by meteors and comets, a theory that was termed 'panspermia' in 1907 by the Nobel-prizewinning Swedish chemist Svante Arrhenius. He actually took the term from Athanasius Kircher who wrote of *panspermia rerum*, 'the universal seed of things'. In turn, he had developed the concept from Bruno's *spermia rerum*, meaning the basic unit of which everything is made – essentially atoms.[5]

Panspermia's most (in)famous recent champions were Sir Fred Hoyle and his long-time collaborator Chandra Wickramasinghe. With a typically robust side-swipe at his peers, Hoyle likened their view that life originated exclusively on Earth to the geocentric ideas that prevailed before Copernicus.[6] In a way he was right, their ideas effectively make our planet the *biological* centre of the universe.

And the increasingly exciting discoveries of the comparatively new field of astrobiology – developed in the late 1950s – reveal that there is no doubt whatsoever that many of the building blocks of life do have an extra-terrestrial origin. The only real controversy is how far they were assembled before they arrived on Earth.

Certainly the chemical ingredients for life exist in space. Even the most remote regions of interstellar space are pervaded with gas and a much, much smaller amount of solid material in the form of extremely fine-grained 'dust'. These cosmic grains are enormously significant. Until the beginning of the 1960s the consensus was that they were simply frozen clumps of gas molecules, but improved technology has revealed that some were too close to stars to be frozen. So what could they be?

Enter the ever-energetic Hoyle and his newly arrived research student from Sri Lanka, Chandra Wickramasinghe. Their time working at Newton's alma mater Trinity College, Cambridge marked the beginning of one of the most enduring scientific collaborations, one that continued after Wickramasinghe's own glittering scientific career took off and only ended with Hoyle's death in 2001.

It was Wickramasinghe who developed the idea that organic carbon-based chemicals form the major components of cosmic dust. When he and Hoyle first proposed this in 1962 it was, unsurprisingly, highly controversial. But research in the 1960s and 1970s vindicated it, and these days it is simply a given.

Formaldehyde, one of the simplest organic compounds, was detected in interstellar clouds in 1969, and since then a whole host of organic chemicals has been added to the list. By the end of the next decade over thirty complex molecules had been found in interstellar dust, including water vapour, carbon monoxide and ammonia. Organic molecules including methane, acids, alcohols and sugars have now been found. Even molecules of vinegar have been detected in a gas cloud in Sagittarius. Around 20 per cent of interstellar dust is now thought to be made up of organic chemicals. The discovery of so many prompted Hoyle and Wickramasinghe to propose, in the mid-1970s, that even more complex organic molecules could be lurking in the interstellar clouds, and that this was a better candidate for the origin of life than the terrestrial 'primordial soup'.

One of the most significant discoveries in this field came in 2005 from a NASA team from the Ames Research Center in California, using data from the Spitzer Space Telescope. The team was studying a type of complex organic molecule with the uncatchy name of polycyclic aromatic hydrocarbons (PAHs), a very common family of chemicals which, in the words of the team's leader Douglas Hudgins, are found 'in every nook and cranny' of the universe. The fact that PAHs are abundant in space had been known for a long time, and few thought they were worth much of a second look. But the NASA team discovered to their great astonishment that the PAHs they were looking at – in a distant galaxy designated M81, 12 million light years away – were rich in nitrogen. This is considerably more significant than it might appear.

Without nitrogen PAHs tend to be hostile to the biochemistry of life. On Earth they are largely the result of the breakdown of organic material, for example the burning of fossil fuels, making them pollutants and in some cases carcinogenic. But with nitrogen it's a different story. Without nitrogen-containing PAHs, amino acids, DNA and RNA, as well as a host of other vital molecules (for example haemoglobin, chlorophyll – and even essentials such as chocolate) could not exist. Indeed, one of the theories of how life originated on Earth puts nitrogen-rich PAHs right at the centre. But the big question is how they developed in the first place.

The discovery that nitrogen-bearing PAHs are present in space provides a major piece of the puzzle. The current understanding, based on the NASA Ames team's work, is that they are formed and ejected into space by the death of stars. As Douglas Hudgins puts it:

There was a time that the assumption was that the origin of life, everything from building simple compounds up to complex life had to happen here on Earth . . .

This stuff contains the building blocks of life, and now we can say they're abundant in space.[7]

Hudgins points out that discovering nitrogen-containing PAHs in interstellar space does not prove that life on Earth came from the stars but that as it is the simplest theory, according to Occam's Razor, this is the one that science should prefer.

ON THE TAIL OF COMETS

Another way that building blocks can be seeded on planets is via comets. Not through scoring a direct hit on Earth – which would incinerate any 'seeds' – but by drifting down with the 'rain' that floats into the atmosphere as the planet passes through the debris from the tail of comets.

Most comets are believed to be left over material from the gas and dust clouds that coalesced at the birth of the solar system, now roaming its highways and byways under the influence of the gravity of the heavenly bodies, generally orbiting the sun. The endless process of heating and freezing as the comet approaches and recedes from the sun causes reactions in its basic chemicals, which creates new compounds.

The idea that seeding occurs via comets has received strong support from the study of meteorites, especially fragments from the famous specimen believed to be from the nucleus of a comet that exploded over the small Australian town of Murchison, Victoria, in 1969. Analysis of the Murchison meteorite has continued ever since – the latest batch of test results, after examination with cutting-edge techniques, was released in February 2010. One of the first things to be discovered was that it was made up of organic carbon chemicals – it even smelt of petrol. It contains 70 different amino acids, including two, glycine and alanine, which are fundamental to life on earth – the very same, in fact, as those that emerged from the Urey-Miller experiments with the primordial soup that so excited scientists back in 1953.

There is an even more direct connection between comets and glycine, which is chemically the simplest amino acid. In 1999 NASA launched the probe Stardust to collect material from the comet Wild-2, which orbits the sun once every six years. In January 2004 Stardust swept up dust from the comet's coma – the cloudy halo around the nucleus – returning it to Earth in a sealed container nearly two years later. Analysis revealed the presence of glycine. As Dr Carl Pilcher, head of NASA's Astrobiology Institute announced:

> The discovery of glycine in a comet supports the idea that the fundamental building blocks of life are prevalent in space, and strengthens the argument that life in the universe may be common rather than rare.[8]

LIFE IN THE LAB

Some of the greatest revelations have come from attempts to reproduce interstellar conditions in the laboratory, in what is effectively a cosmic version of the Urey-Miller experiments. At the forefront of this quest is the NASA Ames Research Center in the 1990s with a team led by Louis Allamandola.

Allamandola set out to study how dust grains in gas clouds interact with the gases by replicating the conditions. He and his team placed methane, water vapour, ammonia, carbon monoxide and so on in extremely thin densities and very, very cold temperatures and bathed them in ultraviolet radiation. Under these conditions, chemical reactions occur that would be impossible under normal earthly conditions. The radiation breaks apart the molecules, and the icy cold puts them back together in very unusual and complex ways. But although many of these structures had never been seen before, some were eerily familiar to biochemists . . .

The first surprises related to the PAHs. Interstellar conditions, particularly exposure to ultraviolet, transform the PAHs' carbon into useful forms for life such as alkaloids – 'ubiquitous in the plant world'[9] – and quinones, essential for photosynthesis and the functioning of muscle and brain cells. These vital substances simply wouldn't exist without the gas and dust clouds in deep space. But there were even more groundbreaking discoveries.

Back in 1985, American biologist David Deamer had discovered something very odd in the Murchison meteorite: small 'bubbles' closely resembling biological structures known as vesicles – basically membrane sacs containing liquid biochemicals that constitute part of cells. But were they really vesicles? It was only in the late 1990s that Deamer realized the potential of Allamandola's work: Could it be that similar bubbles had appeared in his simulated interstellar environment? Indeed they had. They found identical vesicles, about the size of red blood cells. They called in biochemist Jason Dworkin – a former collaborator with Stanley Miller – who identified them as lipids, a class of macromolecule that includes fats and waxes.

Lipids perform several vital functions, but most tantalizingly they make up the membranes of cell walls, which may be small but are in fact big operators. They separate biochemicals into packets – capsules, basically. Without them many of the processes and reactions vital to the development of life could never happen because the biochemicals would be too dilute. As geneticist Pascale Ehrenfreund, specifically commenting on the Ames discovery, pointed out, 'membrane formation is a crucial step to the first forms of life.'[10]

Researchers trying to replicate the origins of life had never previously had any success in replicating lipids under terrestrial conditions. But here, in interstellar gas clouds, they appeared utterly spontaneously – in fact, the Ames team never noticed they had made lipids until David

Deamer asked them to look. The similarity with the vesicles found in the Murchison meteorite shows that this process really does happen in space. It isn't just the Frankenstein child of jiggery pokery in a lab.

In other experiments, Allamandola's team demonstrated that not merely cell membranes, but some of their internal biogenic chemicals – ammonia, formaldehyde and even amino acids – can also be made in the interstellar clouds. Allamanodola speculates that the first cells could have come from inside comets – all the ingredients are there, and so is the membrane to neatly wrap them up – although he admits that this theory is untestable. [11] At least at the moment.

In fact, Allamanodola wasn't the first to make such a suggestion. With so many of life's building blocks being found in space, Fred Hoyle and Chandra Wickramasinghe had suggested back in the 1970s that they could be developed into what they termed 'protocells' that could then be seeded onto planets by comets. The Ames discovery shows this is almost certainly correct.

To us the important thing, though, was the implications of creating lipids in the lab. As Louis Allamandola pointed out:

> The most amazing thing is that we start with something really simple. And then suddenly we're making this enormous range of complex molecules. When I see this kind of complexity forming under these extreme conditions, I begin to really believe that life is a cosmic imperative.[12]

THE LIVING, BREATHING EARTH

The universe – not just planets, but space itself – is bursting with the potential and materials for life, created and transported around like ocean currents carrying seeds between remote islands. However, it is still only on planets that these can develop into something more complex than bacteria at best, which brings us to another even more controversial idea that fits very neatly into the 'universe designed for life' vision.

The Gaia hypothesis was proposed in the 1970s by British scientist James Lovelock, who is even more of a maverick or independent thinker, depending on your point of view, than Sir Fred. Similarly, Lovelock's brilliance is acknowledged even by his critics (even if they believe his imagination sometimes gets the better of him), as are his very real contributions to science.

Lovelock describes himself as an 'independent scientist', neatly encompassing both his attitude to the freedom of thought he believes is essential for a scientist and his avoidance throughout his long career of being tempted by commercial or even academic institutions – although he has occasionally been successfully headhunted as a consultant. With his broad knowledge of all the sciences, and contempt for the increasing specialization that blinkers scientific thinking, he would have been at home in the Renaissance.

Lovelock is that rare being, someone whose brilliance has actually changed the face of the world. Most significantly, this occurred in the early 1970s through his discovery that human-made chlorofluorocarbons (CFCs) had penetrated the environment to such a degree they were present in places as remote from human industry as Antarctica. It was because of Lovelock that today's world is CFC-free.

The concept behind Gaia was a spin-off of Lovelock's work for NASA in the 1960s, when he was devising ways to detect life on Mars. He reckoned analysis of the Martian atmosphere might reveal the characteristic changes caused by the presence of living organisms. Looking further into this question and the impact life has on the Earth's atmosphere led him to certain striking observations. It isn't just that the presence of plants and animals – the biosphere – changes the atmosphere, but they appear to be regulating it, actively keeping the Earth habitable. Life itself keeps the planet in a condition suitable for life.

From such phenomena Lovelock developed the idea that the Earth is a 'self-regulating entity', where living things are not passive guests but ensemble players with integral parts in shaping conditions on the planet.

A prime example of this self-regulation, besides a host of others, relates to the Earth's response to changes in the sun's output. Living organisms can only survive within a narrow range of temperatures, about 10 to 20 degrees centigrade. However, although astrophysicists agree that since life first appeared on Earth at least 3.5 billion years ago the sun's heat has increased by about 30 per cent, the Earth has obviously remained at a temperature suitable for life. Somehow the increasing heat has been balanced to keep the average global temperature steady.

As a 2 per cent drop in the heat reaching Earth from the sun is enough to trigger an ice age, imagine what the Earth would be like with 30 per cent less heat. When life originated, something – probably a high level of greenhouse gases in the atmosphere – made the Earth significantly warmer than it would have been otherwise. But as the sun grew hotter, some other factor must have altered conditions – the mix of gases in the atmosphere, for example – as compensation. And that unknown factor had to keep step with the steady increase in solar heat.

As Lovelock pointed out, any of the processes that have been proposed to explain this compensation would have had to be staggeringly precise. Even small variations in, say, the mix of atmospheric gases would result in runaway reactions that would either seriously overheat the Earth (the oceans would literally boil away), or reduce it to a frozen ball. Yet that clearly didn't happen; the process seems somehow to have been controlled.

> . . . the Earth's living matter, air, oceans, and land surface form a complex system which can be seen as a single organism and which has the capacity to keep our planet a fit place for life.[13]

Following the suggestion of his neighbour, the novelist William Golding, Lovelock called this the 'Gaia hypothesis', after the ancient Greek Earth goddess. In 1979 he produced *Gaia: A New Look at Life on Earth*.

When *Gaia* was first published there were howls of outrage from the scientific world, led predictably by Richard Dawkins. (Lovelock declared that he 'hated Gaia as much as he hates God'.)[14] Dawkins condemned Lovelock's system because to him it could never have evolved by natural selection, while Lovelock maintains that it fits natural selection perfectly. However, as a 2010 BBC documentary on Lovelock's work showed, much of the thinking behind the once-controversial Gaia hypothesis has now become mainstream, while some still regard Lovelock's idea as oddball and over-imaginative, others, including the philosopher John Gray, consider the idea as revolutionary as Charles Darwin's.[15]

Despite widespread belief to the contrary, what the Gaia hypothesis does *not* proclaim is that the world is alive in the same way that an animal is alive, or that it is somehow self-aware, with some higher planetary consciousness controlling and ordering the individual parts to benefit the whole. In fact, Lovelock is scathing about the New Age, which took (or most probably, hugged) his book to its heart, believing in some way that it was scientific proof of the reality of the Mother Goddess. To Lovelock such concepts exist outside the realm of science, since they can't be tested by scientific methods.[16] Lovelock uses the term 'alive' metaphorically, taking pains to explain - with superb chutzpah, but pinpoint precision – 'the planet is alive in the same way that a gene is selfish'.[17] To him the Earth fits the definition of a 'superorganism': 'bounded systems made up partly from living organisms and partly from non-living structural material'[18] – as, for example, a beehive. Literal interpretations of the word 'alive' also seems to be behind Dawkins' hostility, which implies a certain lack of sophistication in his understanding or perhaps an unwillingness to confront the theory properly.

The Gaia theory is what one would expect, indeed predict, from the designer universe hypothesis. If the universe is fine-tuned to support life, and life is a cosmic imperative arising wherever conditions are conducive, then one would expect that once complex life did take hold on a planet, some kind of mechanism would be in place to ensure its survival. We should recall here the concept of the *anima mundi*, the 'world soul', which animates and also controls the world so dear to the Hermeticists' hearts.

But however exciting the Gaia hypothesis might seem, we should remember it has yet to be conclusively proven. Nor does it prove the designer universe theory correct. But, as with the existence of biochemical evidence in support of the cosmic imperative, it nevertheless fits and supports the concept of a designer universe.

THE COSMIC IMPERATIVE

It no longer seems a question of whether panspermia happens – it does, quite clearly – but rather of how far the building blocks of life can be fused in space before they need a planet to really get going. Christian de Duve sums up the current state of our knowledge:

There is . . . ample evidence that a number of biogenic

compounds can form spontaneously under primitive Earth conditions, in interstellar space, and on comets and meteorites. Most likely, such compounds provided the first seeds of life. How much was made locally, how much was brought in from outer space, is still widely debated.[19]

The latest scientific thinking about the origin of life in the universe is very compatible with the concept of a designer universe. Rather than life being a billion-to-one fluke, it seems to be a common – even a *universal* – phenomenon. And the different parts of the universe play vital roles in the creation and dissemination of life.

We must be careful, however, not to put words into the mouths of the likes of Louis Allamandola and Christian de Duve. When they use the expression 'life is a cosmic imperative', they are saying that conditions in the universe mean that wherever life can evolve, it inevitably will. This is emphatically not the same as saying that the 'purpose' of the universe is to produce living organisms. Scientific objectivity and a strict adherence to current evidence could never allow them to draw such a conclusion. But if the universe *is* designed for life, would we be able to tell the difference between that more Hermetic kind of cosmic imperative and de Duve and Allamandola's version?

It is unlikely. If the universe is fine-tuned to produce the chemical elements and right physical conditions for intelligent life, then that same delicate balancing act would have to also include the imperative that biochemistry is now beginning to recognize. It would be pretty pointless otherwise.

Unlike many in the biological sciences, de Duve does give houseroom to the more metaphysical interpretations of the cosmic imperative. In *Vital Dust* he discusses the ideas of Pierre Teilhard de Chardin, the French Jesuit priest and palaeontologist who put forward a theory of cosmic evolution very similar to the designer universe, albeit with a Christian gloss. To Teilhard, creation evolves from simple matter, to life and on to consciousness as part of a divine plan, which de Duve considers a valid possibility.[20]

As biochemistry has become increasingly sophisticated, it has found nothing to contradict the idea of intelligent design. Quite the reverse. However – and to many this will be a very big caveat indeed – the evidence for life as a cosmic imperative is, like that for the fine-tuning of the big bang, hard to square with the image of the biblical God. This is far too limiting for that kind of personal entity, with his alleged omniscience but all-too-human emotions.

An alternative to scientific atheism, which also fits this evidence, is the Hermetic interpretation, in which the cosmos was specifically built for life. Some of de Duve's statements even read like an expression of Hermetic cosmology – a belief in the living universe – albeit in biochemical terms:

> The universe is not the inert cosmos of the physicists, with a little life added for good measure. The universe *is* life, with the necessary infrastructure around; it consists foremost of trillions of biospheres generated and sustained by the rest of the universe . . .
>
> The entire cloud of vital dust forms a huge cosmic laboratory in which life has been experimenting for billions of years.[21]

To de Duve most of the universe exists simply to provide the scaffolding to support life. In his view, the universe is effectively a super-organism in much the same way that, according to James Lovelock, the Earth is. Just on an unimaginably vaster scale.

But what about evolution? Surely the current understanding points in the opposite direction to ideas about life as a universal imperative, or inevitability, good as they may sound. The development of life, especially into anything more complex than a bacterium, is, we are told purely down to chance. If the evolution of life is dependent on random factors, then the idea of design in the universe as a whole is instantly and completely undermined.

Evolution is so often presented to the general public by the Dawkins school as the final coup de grâce, not merely to religious creationism, but also to any idea of design behind the universe, that it may seem perverse even to begin to challenge it. But what happens if you dare do just that? The results are rather surprising, although they won't turn you into a creationist. In fact, quite the reverse. As we are about to see, the theory of evolution so beloved of Dawkins et al., by no means proves atheism to be right.

CHAPTER ELEVEN

DARWIN'S NEW CLOTHES

Today's only accepted and acceptable scientific theories about the origins and development of living things reject even the slightest degree of design. Instead, the whole process that has fashioned the dazzling display of animals, plants and micro-organisms that cover the Earth is, we are told, driven ultimately by blind chance.

Evolution has become the really big battleground for the righteous – or perhaps, more accurately, the self-righteous – in the conflict between science and religion, particularly between militant atheists and Christian fundamentalists.

For those who take Genesis literally, evolutionary theory has not only to be rejected but also actively anathematized. The first book of the Bible states that God made all plants, sea creatures, birds and land animals (in that order) 'according to their kinds' – as individual, and by implication, fixed species. If, as science now understands it, different species developed one from another, then the biblical account is basically just wrong. Even worse to the Genesis literalists is the notion that humans – to whose creation God is supposed to have devoted special care and attention, making us 'in his own image', no less – are part of that scheme, that we have evolved from lower animals.

But scientists have made evolution a battleground too, seeing it as their greatest victory over the forces of superstition and irrationality, and raising the fear that undermining it will see the end of their intellectual triumph. In the last couple of decades there have been good reasons for scientists to be anxious, as the recent political controversy in the USA over intelligent design (ID) has shown. The well-organized and generously-funded ID movement aims to undermine evolutionary theory by picking on its flaws, but it does so as part of a Christian fundamentalist – creationist – agenda. So, if biologists admit that the theory is anything less than cast-iron, their opponents will pounce and, particularly in America, use such admissions for political ends, their immediate objective being control of the education system.

The ID movement emerged as the result of a series of reversals that Christian fundamentalists have suffered since the 1980s, in which attempts to have creationism taught compulsorily in state school science classes were successfully challenged in the Supreme Court. These were ruled unconstitutional because the United States' constitution – its First Amendment, which dates back to 1791 – explicitly separates Church and State.

Creationists then began to recast their argument in more scientific-sounding terms, basically crossing out 'God' and 'creation' and replacing them with 'designer' and 'intelligent design'. The phrase 'intelligent design' was carefully chosen, as it has occasionally cropped up in the scientific literature over the years. Charles Darwin himself used it.

The ID movement's strategy is to highlight apparent gaps in Darwinian theory and biological phenomena that are either hard to explain in Darwinian terms or which seem to actively contradict it. It goes for the weak spots and then offer, intelligent design as an alternative. Of course, it is possible to believe in intelligent design without being a Christian fundamentalist; it's just that virtually all ID-ers are.

But – and this is an important point – many of the ID movement's claims about Darwinism's weaknesses aren't its own, but are lifted from the works of bona fide scientists. The notion that some creative, guiding and purposeful factor influences biological evolution has been proposed by dispassionate and objective thinkers with no religious axe to grind. Indeed, that great proponent of the 'intelligent universe', Sir Fred Hoyle, could have given Richard Dawkins a run for his money in the anti-organized religion stakes. The ID movement is cynically twisting such challenges to serve its own agenda.

Given such resolute opponents, small wonder that the scientific community sees any attempt to challenge Darwinian orthodoxy as dangerous and religiously motivated. Anyone who argues against it is assumed to be hiding a creationist agenda. This makes the whole subject a minefield for those who fully intend to get to the bottom of the subject, no matter where it might lead.

Of course there will be many who disapprove of non-specialists investigating the subject in the first place. But often those who devote decades to one aspect of a complex discipline end up simply not being able to see the wood for the trees. We, on the other hand, can stand back and see the wider picture. One way of doing so – especially where academic sacred cows like evolution are concerned – is to revert to childhood. One specific, fictional childhood in particular will provide some much-needed perspective. We are assuming the role of the little boy in Hans Christian Andersen's fable, the lone critic of the Emperor's 'new clothes'. In this tale, everyone agreed they were magnificent – except for the young outsider who saw that they were, in fact, completely non-existent. Following his lead, we also find ourselves standing towards the back of the crowd, ignoring the cheering to see what is really there.

CHANCE WOULD BE A FINE THING

Famously, the cornerstone of evolutionary theory is natural selection, or survival of the fittest, as proposed by Charles Darwin (1809–82), most prominently in his *On the Origin of Species by Means of Natural Selection* in 1859, which has since become the bible of modern biology. Actually 'survival of the fittest' was coined by the philosopher Herbert Spencer as a way of avoiding the implication of design in the phrase 'natural *selection*'. Even back then people were cautious about giving ammunition to creationists.

In Richard Dawkins' hands, natural selection has been moulded into a quasi-religious revelation. To him natural selection also achieves a very rare thing: it proves a negative by showing that God does not exist. Natural selection provided Dawkins with his atheist epiphany, as well as being the catalyst that 'raised his consciousness',[1] to use one of his favourite phrases.

To Darwin's natural selection, modern biology has added genetics, the mechanism of heredity first proposed by – another irony – a Catholic monk, Gregor Mendel, in 1865 (although the term 'gene' was only coined in the first decade of the twentieth century), and since the discoveries of Francis Crick and James D. Watson in the 1950s known to operate through DNA. 'Neo-Darwinian theory' or 'Neo-Darwinian synthesis' is basically natural selection plus genetics.

The basic principles of natural selection are familiar and straightforward enough. If a new trait appears in an individual animal or plant that gives it an edge in the survival game – helping it be more efficient at finding food, dodging predators or attracting a mate – it will out-perform the rest of its species. It will live longer and produce more offspring that, inheriting the new trait, will also be one step ahead in the survival stakes. Eventually, after many generations, only those with the new feature will remain, the species having evolved into something new and better. Many more generations later, it will have become a new species entirely, unable to breed with members of the 'parent' species. Conversely any new traits that hamper an organism's ability to survive or reproduce will be self-evidently eliminated.

As to what causes the changes on which natural selection works, it is all down to changes in DNA. When this miraculous molecule replicates during cell division, it nearly always reproduces itself perfectly. Extremely rarely a change is introduced, and when this happens it changes something in the organism's physical form or in one of the biochemical processes that sustains it.

Changes in even a single gene can have the most profound effect. One mutation, for example, results in a mammal's hind legs remaining vestigial within the body. Although an animal with such a handicap wouldn't last very long, there are rare situations in which the mutation can actually be useful: it would help streamline semi-aquatic mammals, for example. In fact, fully aquatic whales and dolphins have been shown to have exactly that mutation.

Natural selection is not the driving force of evolution; genetic mutation is. Natural selection is more of a steering force, either gifting a change to the rest of a species or simply eliminating it. But what causes the mutations? According to the consensus, they arise from random and unpredictable copying errors that occur during replication. So, although the genetic system is beautifully elegant, life in all its myriad forms owes its existence to the imperfections in this system.

The process of random genetic mutation and natural selection, we are told, accounts for all of the enormous diversity of life on Earth. Everything that lives – microbial, animal or vegetable – has evolved over the course of billions of years from a single original organism, the 'cenancestor'. (Otherwise known as the more zappy LUCA, 'Last Universal Common Ancestor', a term presumably chosen because the more apt 'First Universal Common Ancestor' would result in a somewhat inappropriate acronym.)

As Francis Crick wrote (his emphasis), *'Chance is the only true source of novelty'*.[2] Similarly, in the 1960s the Nobel-prize winning biologist Jacques Monod uncompromisingly put man in his place. With typically French existential angst he wrote:

> The ancient covenant is in pieces; man at last knows that he is alone in the unfeeling immensity of the universe, out of which he emerged only by chance. Neither his destiny nor his duty have been written down. The kingdom above or the darkness below: it is for him to choose.[3]

Put so depressingly, it's hard to see much of a choice.

But does chance alone really explain everything in the natural world? Copying errors do happen – as is proven by genetic disorders – but if every individual tweak to the genetic code is random, can they alone explain all the vast number of changes needed to transform LUCA into human beings, E. coli, broccoli, whales and duck-billed platypuses?

Introducing errors into any system isn't usually a clever idea. Fred Hoyle and Chandra Wickramasinghe astutely observed that ascribing all the variety of the animal, vegetable and microbial worlds to random mutations is like saying a computer program can be improved by intro-ducing random mistakes.[4] And Paul Davies writes in *The Cosmic Blueprint* (1988) that logically:

> one would suppose that random mutations in biology would tend to downgrade, rather than enhance, the

complex and intricate adaptedness of an organism. This is indeed the case, as direct experiment has shown: most mutations are harmful.[5]

The standard response to this is that the vast majority of DNA mutations are indeed harmful, but natural selection weeds them out by killing off the afflicted animal or plant. The number of mutations that just happen to be beneficial might be minuscule, but they are enough, we're told confidently, to account for everything that ever evolved. But this is by no means solid fact: it is actually just an assumption.

The problem for evolutionary scientists is that the factors involved are impossible to quantify. Mutations during DNA replication are extremely rare. According to John Maynard Smith, one of the late twentieth century's foremost geneticists, each time DNA replicates, the chance of a change in a base pair is one in a thousand million.[6] Most mutations have no effect anyway because the genetic system has a clever error-correcting mechanism. And the vast majority of mutations that do have an effect on the individual organism make no difference in evolutionary terms. The only changes passed on to the next generation are those which happen in the 'germ line' cells – sperm and eggs and the cells from which they develop. Only a minute percentage of *those* produce a beneficial change in the organism; most do damage. It is impossible to put precise figures on any of this.

The other side of the equation involves the speed of evolution, or how long it takes one particular species to evolve from another, which entails identifying the genetic changes responsible. As evolutionists can rarely, if ever, determine either of these with anything approaching certainty, there is ultimately no way they can prove that chance and chance alone was responsible.

Evolution is dependent on so many factors – the appearance of 'good' mutations, the size of a species' population, competition from other animals, its environment and the speed of environmental changes to which it has to adapt or die. The origin of each species, every branch in the evolutionary tree, is a special, unique case, as Francis Crick asserts:

> Strictly speaking, we can form no firmer estimate about the time needed for evolution than we can for the chance of any particular step . . . There is no detailed theory of evolution so quantitative that we can calculate just how long any particular stage is likely to require.[7]

Ever since Darwin, the physical changes on which natural selection works have been *assumed* to be purely random. The reason is obvious: if these changes aren't the result of chance alone, then some other factor or factors are responsible, and there is no conceivable way to account for such factors without invoking the supernatural.

In order to make this assumption work, evolutionary theory relies on an egregiously circular argument, which basically goes as follows: No matter how unlikely it seems that a particular characteristic should evolve through random mutations, it must have done, because it now exists – and only random mutations can make things evolve. Frankly, this is outrageous. If non-Darwinists used similar (non) logic, we would be hammered – and quite right, too.

To be fair to evolutionary biologists, their inability to prove the quintessential importance of chance does not necessarily mean the theory is wrong. There are, however, many events in evolutionary history that are not merely difficult, but impossible to explain in neo-Darwinian terms. In fact, astoundingly, *most* of the major steps in the advancement of life, from the primeval to the complex, fall into this category. Even mainstream biology acknowledges that processes outside the normal neo-Darwinian mechanism are required for these steps, or else pronounces itself completely baffled.

THE GREAT DNA MYSTERY

The first big mystery is how DNA itself came into being. After all, the entire variation of life on earth is essentially the result of the shuffling and reshuffling of its basic code. As one researcher put it recently, DNA 'has multiplied itself into an incalculable number of species, while remaining exactly the same'.[8]

There is a fundamental Catch-22 situation about the origin of life. In order to replicate, DNA requires certain proteins in the form of enzymes to act as a catalyst, but no protein can be produced without DNA in the first place. At present, there are only theories that seek to explain how this came about, which because of their very nature are untestable.

In the mid-1980s a suggestion by British molecular biologist Graham Cairns-Smith that the earliest 'genes' evolved from clays attracted considerable interest. A current favourite is the 'RNA world' theory, which proposes that in the early stages, when only primitive single-celled organisms existed, life was based on RNA rather than DNA. We also mentioned earlier the PAH world hypothesis, according to which polycyclic aromatic hydrocarbons once predominated, leading to the development of RNA. However, although it makes sense that PAHs came first, were followed by RNA and then DNA, this theory is also rather vague.

All of these hypotheses, naturally, assume that the process of development from ordinary chemicals to the fully-fledged genetic system was entirely due to blind chemical reactions and chance. But that's just an assumption, and it gets worse: there are only the vaguest ideas about exactly how this happened. As Christian de Duve comments in *Life Evolving* (2002):

> . . . we are mostly left with speculative hypotheses to explain the manner in which the basic building blocks provided by cosmic chemistry might have combined into larger molecules, such as proteins and, especially, nucleic acids, not counting the more complex assemblages from which the first biological structures arose. One may well wonder, therefore, whether we will ever succeed in explaining the origin of life naturally or, even, whether this phenomenon is naturally explainable.[9]

Life, and therefore DNA, appears to have been here almost as soon as the planet had reached the right conditions. There seems suspiciously little time for it to have evolved through random events.

And there is a further twist: DNA seems to have evolved twice. Until the 1970s it was thought that all life could be divided into two 'domains', depending on their type of cell. These were bacteria and the more complex 'eukaryotes' – everything that isn't bacteria, including all the really complex stuff such as animals and plants. Basically the eukaryotic cell has a nucleus, whereas the bacterial cell doesn't.

Then in 1977 American microbiologist Carl Woese made an apple-cart-upsetting discovery at the University of Illinois. It turned out that some 'bacteria' were actually something else entirely. Although these organisms were, like bacteria, single-cell microbes without nuclei, they are as genetically distinct from bacteria as bacteria are from eukaryotes. Woese named this new, third type of organism archaea, from the Greek meaning 'beginning' or 'primeval'.

Unexpectedly, molecular biologists discovered that bacteria use different enzymes to replicate their DNA from those used by eukaryotes and archaea – revealing that there are two entirely different systems of DNA replication.[10] Since DNA controls its own replication, this means there are two quite separate and independent types of DNA. Basically, as geneticist Anthony Poole of Stockholm University noted: 'What it really looks like is that DNA has evolved twice.'[11] There was therefore not one but two LUCAs, one the ancestor of bacteria, and the other of everything else. Assuming it is all due to chance, something with extraordinarily long odds actually happened twice – both times very early in the Earth's history – and never happened again.

Some scientists, such as Carl Woese, now acknowledge that it is impossible to explain the evolution of the genetic code in purely Darwinian terms, and are exploring alternative mechanisms for the origin of DNA.[12]

So nobody knows. Not even a little bit. All the ideas put forward are still too clunky to count. Leading palaeontologist Simon Conway Morris laments that scientists' inability to discover the origin of life is 'one of the great scientific failures of the last fifty years'.[13] Even Dawkins stays out of the mix, but only, he is keen to point out, because the search for the origin of life, being a question of chemistry, is outside his field of expertise.[14] It's frustrating and sobering to realize that although we know what must have happened for life to get started, we haven't the faintest idea how. It certainly suggests that evolutionists who declare dogmatically that the origin of life owes nothing to non-random factors are vaingloriously jumping the gun. They just can't be sure.

THE BIG ANAL BREAKTHROUGH

Although DNA is the prerequisite for life, there are other key milestones in the journey from single-celled microbes to today's complex life forms. And without these, no further progress up the evolutionary tree could ever have been possible.

Many of these landmark events are obvious, such as the development of vertebrae, but some are more unexpected, including the appearance of the anus, sometimes called somewhat eye-wateringly the 'anal breakthrough', which apparently occurred some 550 million years ago. Without an anus, mouths couldn't evolve – or if they did without benefit of a rectum, animals would explode after a couple of meals – and without mouths heads couldn't evolve, and without heads we couldn't have sizeable brains. This prompted one of our favourite quotes in evolutionary literature, from Oxford zoologist Thomas Cavalier-Smith: 'The anus was a prerequisite for intelligence.'[15] (Given the pronouncements of certain dogmatists, we always suspected as much.)

Another of life's most vital developments was the appearance, some two billion years ago, of a revolutionary new type of cell, the complex and large eukaryotic cell that we mentioned above. Before its appearance there was only the more primitive bacterial, 'prokaryotic' cell.[16] The crucial difference between the two is that the eukaryotic cell has a DNA-filled nucleus, whereas in the prokaryotic the DNA is diffused throughout the cell. Eukaryotic cells have up to one thousand times more DNA. And the nucleus arrangement means that only eukaryotic cells can develop into large, more complex organisms – all animals and plants are eukaryotes. Without this type of cell, the Earth would still be populated exclusively by microbes. Microbes without anuses, that is.

Logically the eukaryotic cell must have evolved from the

simpler prokaryotic. As Cavalier-Smith notes, this process 'involved the most radical changes in cell structure and division mechanism in the history of life'.[17] He adds that the leap required 'dramatically accelerated evolutionary rates for many genes and, more importantly, massive novel gene creation'.[18] But as leading cell biologist Lynn Margulis, writing with her son Dorion Sagan, acknowledges:

> The biological transition between bacteria and nucleated cell, that is between prokaryotes and eukaryotes, is so sudden it cannot effectively be explained by gradual changes over time. The division between bacteria and the new cells is, in fact, the most dramatic in all biology.[19]

They go on to explain:

> All cells either have a nucleus or do not. No intermediates exist. The abruptness of their appearance in the fossil record, the total discontinuity between living forms with and without nuclei, and the puzzling complexity of internal self-reproducing organelles suggest that the new cells were begotten by a process fundamentally different from simple mutation or bacterial genetic transfer.[20]

In other words, this vital leap simply *cannot* be explained by the usual neo-Darwinian chance mutation and natural selection. There has to be a completely different process involved.

Lynn Margulis' groundbreaking solution to the conundrum, proposed in the mid-1960s and which revolutionized the understanding of cells (after the usual years of disparagement and dismissal from her peers), was that it was an act of symbiosis: some kinds of prokaryotic cells entered others, feeding off their waste products and leaving their own detritus as food.

But even this only explains *what* happened. We are no nearer to knowing how or why. Evolutionary scientists fully accept that such a transition required a special, unique – and resolutely non neo-Darwinian – set of processes, without which multi-celled life could never have existed.

THE IMPOSSIBILITY OF SEX

Another milestone after eukaryotes was the development of sexual reproduction, without which no complex life would be possible – and something else about which evolutionary biologists tie themselves in knots. Metaphorically, at least.

The simplest micro-organisms reproduce asexually, by splitting into two, each half containing the same DNA. From the first appearance of life around 3.5 to 4 billion years ago until, according to the evidence of microfossils, between a billion and a billion and a half years ago, that was the only kind of reproduction there was. As each new cell is essentially a clone of its 'parent' – they are genetically identical, the DNA being passed on unchanged – it doesn't allow for much genetic diversity, making evolution very slow, which is why not much happened for some three billion years.

Sex is by far the better option for the evolution of more complex and intelligent organisms. Genes are packaged in chromosomes, and during reproduction those from each parent are split up and then recombined. No new genes are created – that's still down to mutation – but new *combinations* of genes are thrown up. The process of recombination creates genetic diversity in a way that asexual reproduction never can. Natural selection has more options to try out. It also allows beneficial mutations to spread throughout a species more easily – basically speeding up evolution.

The evolution of sex is another of biology's great unsolved riddles. It is easy enough to see *why* it happened, but it has proved impossible to work out *how*. The leading American evolutionary biologist George C. Williams wrote that sex is 'the outstanding puzzle in evolutionary biology'.[21] His *Sex and Evolution* (1975) opens with the sentence: 'This book is written from a conviction that the prevalence of sexual reproduction in higher plants and animals is inconsistent with current evolutionary theory.'[22] His conclusions have a somewhat forlorn tone:

> I am sure that many readers have already concluded that I really do not understand the role of sex in either organic or biotic evolution. At least I can claim, on the basis of the conflicting views in the recent literature, the consolation of abundant company.[23]

John Maynard Smith also devoted a volume, *The Evolution of Sex* (1978), to the various theories on the subject. He, too, concluded forlornly: 'I fear that the reader may find these models insubstantial and unsatisfactory. But they are the best we have.'[24] In a later essay entitled 'Why Sex?', Smith says that even though he has devoted twenty years to the problem of sexual evolution, 'I am not sure I know the answer.'[25] (Even so, he still received the Royal Society's Darwin Award for his contributions to research on the evolution of sex. It seems a little unfair on the rest of us who also don't know the answer.)

Little progress has been made since the 1970s. In *Evolution* (2004), zoologist and science writer Matt Ridley examines all the most popular theories about how sex evolved, and after finding major problems with the lot, concludes that 'the existence of sex is the profoundest puzzle of all'.[26]

The prevailing theory is that sex began with the chance fusing of cells infected with different but very similar viruses. When the cells divided, differences between the viruses resulted in a replication of the DNA that prefigured the workings of chromosomes. If this theory is correct, this vitally important change without which nothing bigger than a virus could exist, didn't even involve a genetic mutation. Even more than other evolutionary changes, it was pure fluke. Like the all-important appearance of eukaryotes discussed above, sex is another thing that owes nothing to the usual neo-Darwinian mechanism.

John Maynard Smith makes the major point that although we think of sex and reproduction as the same, genetically speaking they're the exact opposite. Reproduction turns one cell into two, while sex fuses two to make one. He goes on:

> Darwin has taught us to expect organisms to have properties that ensure successful survival and repro-duction. Why, then, should they bother with sex, which interrupts reproduction? . . . It is not merely that sex seems pointless: it is actually costly.[27]

The big cost is the necessity of producing and maintaining males. Compared to asexual reproduction, sex takes twice as many organisms to produce the same number of offspring. Fewer offspring are produced and more slowly. These weren't obstacles once sex caught on, but would have been severely restricting in the very earliest stages, when the primitive sexual organisms were in competition with the asexuals, which should have out-bred them. As Williams comments: 'This immediate advantage of asexual reproduction is generally conceded by those who have seriously concerned themselves with the problem.'[28]

As everyone knows from experience, sexually repro-ducing animals have to devote time and energy to finding mates that could be better used ensuring their survival. And we see the palaver it causes just in the animal and bird world when even after all that strutting and rutting and preening, it is still possible to get rejected – or eaten. As Lynn Margulis and Dorion Sagan acknowledge: 'Biologically, sexual reproduction is still a waste of energy and time.'[29] Many would agree.

For the individual organism, asexual breeding is much better, requiring less energy and biochemical complication. And, according to neo-Darwinism, the individual level is all that matters. The fact that doing things differently might be better for the species as a whole, or for the progress of life in general, is irrelevant. New systems are only adopted if they help the individual; helping the species is just an accidental by-product. As John Maynard Smith acknowledges, both obvious advantages – genetic recombination and speedier evolution – ascribe foresight to evolution, which would cause any self-respecting neo-Darwinist to have apoplexy.[30]

In fact, theoretically, sex shouldn't exist, as Williams admits: 'The impossibility of sex being an immediate reproductive adaptation in higher organisms would seem to be as firmly established a conclusion as can be found in current evolutionary thought.'[31]

But the impossible *did* happen. That was lucky.

And as for the big puzzle of why sex was invented, although of course it would be facetious to suggest it is because it's more fun than cell division, frankly that's as good an idea as any other at the present time.

SEX AND DEATH

A similar situation applies to the phenomenon of ageing, common to everything above the simplest organisms in the animal world. In nature, ageing *is* death: in the wild individuals rarely have the luxury of dying of old age, as the inability to run away, fight or even chew food properly carries its own death warrant. Without ageing and death the evolution of ever more complex organisms would be impossible. And yet it is far from clear how ageing evolved.

Strange as it may seem, rather than simply being the result of the body wearing out, the build-up of toxins or accumulated oxidization from free radicals, ageing is due to a genetic switch that halts the repair and regeneration processes at a cellular level. Once the repair mechanisms stop, we start to age. While some individual problems of old age, such as cataracts, are due to the length of time an individual has lived, that's not the same as the general condition of ageing, or senescence. Old age is basically a pre-programmed phase of life, just like puberty. But whereas puberty has an obvious biological function, what on earth is the purpose of ageing?

If ageing is genetic, self-evidently it must have evolved. Indeed, in the 1990s studies of the genomes of different species found it was due to specific genes that are shared throughout the evolutionary tree, from yeast to mammals.[32] An irreversible decline seems to be a common feature of eukaryotes, and emerged at around the same time as sexual reproduction.

The genetic basis of ageing presents something of a problem for natural selection, in which survival is allegedly paramount. To put it kindly, it is a paradox. After all, what's the survival value of something that kills you?

And there's another problem: how did the ageing genes get passed on in the first place? There must have been a point, very early in the life of eukaryotes, when the genes didn't exist. Therefore mutations must have created them. For most of the organism's life, and especially during its most fertile time, those genes would be irrelevant: they only have an effect when the switch is thrown. So why, then, would natural selection favour them? Why would individuals with the mutations be more successful, producing ever more offspring?

Evolutionary biologists are unable to answer these questions. There aren't even many theories. The most popular hypothesis, that of 'antagonistic pleiotropy' put forward in the 1950s by George C. Williams, was blown out of the water in the 1990s by laboratory experiments. Briefly, Williams' theory was based on the idea that the ageing genes must also have *beneficial* effects, especially early in life, and although they may have a deleterious effect later this doesn't matter, since the majority of organisms in the wild seldom live to old age anyway. (Live fast, die young, in other words.) Natural selection favoured individuals with the genes because it gave them early advantages, any later drawbacks being irrelevant. Although this may be the only hypothesis that could explain senescence while remaining dutifully neo-Darwinist, *it doesn't work.* New discoveries have highlighted its drawbacks, such as the ageing genes in yeast, for example. And laboratory experiments have not only failed to prove the theory's predictions but have come up with diametrically opposite results – selectively breeding fruit flies to live longer, for example, has shown them to be fitter in early life, too.[33]

The very few other theories all raise more questions than answers. How ageing evolved is literally another one of life's unsolved mysteries.

There is only one known species that is, quite literally, immortal, barring accidents and disease. This is a tiny, 5 mm hydrozoan, *Turritopsis nutricula* – a sort of jellyfish native to the Caribbean – whose special biological talent was only discovered in 2009. *T. nutricula*'s trick is to revert to its sexually immature stage after reproducing, going through an endless cycle of infancy and adulthood. Although apparently unique, it does demonstrate that immortality can evolve. But why isn't it more common, especially since obviously the ultimate in natural selection would not be mere survival but actual immortality?

As with sex, it's easy to see the advantages that ageing has for a *species*, and for the progress of life in general. It avoids over-population and therefore competition for resources. Just imagine what would happen if a species were both immortal and fertile! It also retains the all-important genetic diversity by renewing the entire population periodically. If older generations didn't die, and were able to mate with younger generations, then a species would never be able to eradicate its old genes. No new, improved genes would ever get a chance to catch on.

It is tempting to speculate that senescence developed specifically in response to the evolution of sex, in order to avoid these problems. Without death, after all, the benefits of sex for the faster spreading of life-improving genes throughout a species would be lost. The only drawback to this neat explanation is that Darwinian theory doesn't allow for it.

The avoidance of overpopulation and the clearing out of the gene pool was, around the turn of the twentieth century, the most popular explanation even among Darwinists for the development of ageing. But it then dawned that this explanation actually contradicts Darwinism as it assumes that the species as a whole somehow knows what is good for it in the long run. Getting rid of the older generations is advantageous to a species as a whole, but can hardly be said to be much good for an individual, and it is changes in the individual that drive evolution. It's another one of those awkward catch-22 situations that make us feel as though we're missing something vital, somewhere.

CREEPS AND JERKS

Almost incredibly, neo-Darwinism also has difficulty in explaining – of all things – the origin of species . . .

The theory of speciation says that when a beneficial mutation occurs, over the course of many generations natural selection carries the new trait to the rest of the species. Eventually so many changes from enough mutant individuals accumulate that a new species comes into being. The new species is genetically distinct from the original, to the point that it cannot breed with any members of the original species that might still be around. Given the air of confidence with which such matters are discussed in the public domain it's surprising that evolutionary biologists can't agree about how the processes governing speciation occur.

Different schools of evolutionary biology have proposed different models of exactly how speciation happens, but none of them can prove theirs to be correct. This is not surprising: it's another of those areas where it is virtually impossible to acquire hard data. Evolution is such a long, slow process. After all, you can't watch it happening in a lab, at least not where the likes of elephants, rubber plants or quantum physicists are concerned. There have been some instructive experiments on the micro level, with bacteria – particularly a long-running series with E. coli at Michigan State University – where different strains have changed their genetic makeup over many generations. However, such experiments involve primitive species in glorious isolation facing limited survival problems, which hardly mimics the conditions of the real world.

All the scientists really have to work with is observation of the natural world and analysis of fossils, both of which are severely restricted. Contrary to expectations, the fossil record is not much help for the neo-Darwinian model, since many things that the theory would predict to be there are conspicuous by their absence. Darwin himself, in the words of palaeontologist Stephen Jay Gould, 'viewed palaeontology more as an embarrassment than as an aid to his theory'.[34] And eminent neo-Darwinian Ernst Mayr acknowledged in the late 1980s that following the fossil record:

... seemed to reveal only minimal gradual changes but no clear evidence for any change of a species into a different genus or for the gradual evolution of an evolutionary novelty. Anything truly novel always seemed to appear quite abruptly in the fossil record.[35]

More recently, Steve Jones, Professor of Genetics at University College, London, has stated that the fossil record 'can look anti-Darwinian', meaning many of the things that should be there just aren't.[36]

These absences are contrary to all Darwinian expectations. After all, the most dramatic changes should take the most time to manifest, and should leave more fossil traces. Yet they are not there. It has to be assumed that this anomaly is due to the fragmentary nature of the fossil record.

True, we are left with the remains of a tiny fraction of all the animals and plants that ever lived, representing a tiny fraction of all the species that have ever evolved. Only about a quarter of a million different species have been found in fossils, whereas the number of species ever to live is probably up there in the billions. The fossil record is really a random sampling of evolutionary history. How random and how big a sample nobody really knows – palaeontologists are left floundering in a statistical gloom.

Others are not so sure that the leaps in the fossil record can be brushed aside quite so easily. Although most biologists maintain Darwin's original view that evolution is a slow and gradual process, for a minority in the field – mostly palaeontologists – it happens in short, sharp bursts. These are the theories of quantum evolution proposed by George C. Simpson in the 1940s (which still has supporters such as Thomas Cavalier-Smith), and punctuated equilibrium put forward by Stephen Jay Gould and Niles Eldredge in the early 1970s. While the two theories agree the story of each species consists of long periods of stasis with short bursts of rapid evolution, their proposed mechanisms are quite different.

Critics of punctuated equilibrium have called it a theory of evolution by jerks; Gould responded that theirs is a theory of evolution by creeps. But punctuated equilibrium and quantum evolution do at least suggest why the fossil record offers scant evidence of the gradual metamorphosis of one species into another.

The major objection to quantum evolution and punctuated equilibrium is that they require a mechanism over and above 'classic' neo-Darwinism, implying that the current theory is incomplete – hardly music to the ears of most evolutionists.[37] Other biologists argue the theory is missing something vital. British biologist Brian Goodwin declares:

> . . . despite the power of molecular genetics to reveal the hereditary essences of organisms, the large-scale aspects of evolution remain unexplained, including the origin of species . . . So Darwin's assumption that the tree of life is a consequence of the gradual accumulation of small hereditary differences appears to be without significant support. Some other process is responsible for the emergent properties of life, those distinctive features that separate one group of organisms from another – fishes and amphibians, worms and insects, horsetails and grasses. Clearly something is missing from biology.[38]

'LOOK AT THE KING! LOOK AT THE KING!'

Most of the problems highlighted above would be solved if there were some way that natural selection could operate at the level of species, or even higher. Something that could see the big picture, in other words. But there is no place in neo-Darwinian theory for this. A species evolves because the individuals within it evolve. Natural selection does not work at the level of the species, or the gene, but the individual.[39]

We are told, with confidence edging into arrogance, that neo-Darwinism can explain everything in the biological world, and there's no need to invoke anything else. However, as we have seen, it totally fails to explain:

- The origin of life itself, specifically the origin of DNA.
- The appearance of the nucleated, eukaryotic cell without which multi-celled life would be impossible. (A 'special case', the result of a process outside the usual neo-Darwinian model.)
- The origin of sexual reproduction, another thing without which complex organisms couldn't evolve. (Another special case that required a non-Darwinian process.) Not to mention how sex caught on, given all its disadvantages.
- How ageing, the clearing out of the gene pool without which evolution couldn't advance, came into being.
- And – irony of ironies – Darwinism can't really explain exactly how species originate.

Frankly, the Emperor is just plain naked. Nude. His only suit is the one he received on his birth day.

No doubt Richard Dawkins will be sighing as – or rather if – he reads this, 'Here we have yet more non-scientists picking holes in Darwinism just because it can't explain everything . . . so far at least . . . ' But there is an elephant in the room that is particularly difficult to miss. In fact, there are so many glaring flaws in the logic of neo-Darwinism that there is a whole herd of deliberately unnoticed pachyderms crammed into that one little space.

Darwinism performs a neat sleight of hand by using

observations as explanations. Although perhaps an over-simplification, there is nevertheless some truth in the way that the great iconoclast of scientific theorizing and collector of strange phenomena, Charles Fort sums up the evolutionary message: 'survivors survive'.[40] It's not so very different from the logic behind the quip: 'Statistically, people who have the most birthdays live longest.'

Neo-Darwinians do have a penchant for seeking to explain all biological phenomena simply by describing them. Take for example convergent evolution – perhaps 'parallel' might be a more apt term – where two species widely separated on the evolutionary tree have independently developed exactly the same anatomical solutions to the same survival problems, without having inherited them from a common ancestor.

There is a plethora of impressive examples across the animal and plant kingdoms where organisms that look virtually identical are in fact completely unrelated genetically. Many of the most obvious are found in Australia, which because it has been cut off from the other continents for around 50 million years, has developed its own idiosyncratic flora and fauna. In particular, marsupials rule, whereas in the rest of the world mammals with placentas have won the day. This has resulted in many Australian creatures that, fitting the same ecological niche as placental mammals, have evolved a very similar anatomy. There are marsupial moles, which look like moles from elsewhere, marsupial mice that look like non-Australian mice, and even an equivalent of the flying squirrel, the flying phalanger. Since marsupial and placental mammals diverged far back down the evolutionary tree, all of these have evolved completely independently.

But evolutionary theory also recognizes divergent evolution, where different species facing the same survival problems in similar environments come up with *different* solutions. Yet both types of evolution – convergent and divergent – are regularly cited as definitive proof of Darwinism. For example, *New Scientist's* biology features editor Michael Le Page, wrote of divergent evolution in a 2008 article intended to counter the claims of the intelligent design movement, that 'there is no reason why a "designer" would not have mixed up these features'.[41] Dawkins also argues that the absence of shared features in distantly related species is evidence against intelligent design – no mammal has feathers, even though they would be useful to flying mammals such as bats.[42] But convergent evolution is just the kind of mixing up that Le Page says never happens; according to his and Dawkins' logic, convergent evolution must be evidence for design.

Even more bizarrely, Dawkins uses convergent evolution as an argument against intelligent design. He counters creationist claims that complex organs such as the camera eye of mammals – which is made up of separate components that individually do nothing but work perfectly together – could not have evolved by chance. Dawkins points out that the camera eye has actually evolved independently at least seven times (and eyes of any kind, based on other principles, at least forty).[43] Not only is this a non sequitur – surely it just makes the problem seven times worse – but it also contradicts his argument that the lack of shared features in distantly related species disproves the existence of a designer.

Besides divergent and convergent evolution adapting a species to its environment, there is a third option: no evolution at all, or stasis. Judging by the fossil record, some species – 'living fossils' as Darwin put it – have hardly changed over vast tracts of time, including sharks, crocodiles, horseshoe crabs and horsetail plants. According to Ernst Mayr: 'Some species are extraordinarily young, having originated only 2,000 to 10,000 years ago, while others have not changed visibly in 10 to 50 million years.'[44]

But *why* didn't the living fossils change even a tiny bit over such an extraordinary length of time, when most species quite clearly have? Unsurprisingly nobody knows for sure. It is often said that the non-changers are just perfectly adapted to their environment ('evolutionary complacent', as British comedian David Mitchell puts it), but this is simply putting too positive a gloss on it. The evolutionary explanation is rather that it is because these smug species are so finely attuned to their environment that the slightest change means they can't survive in their own little niche, so no changes ever get a chance to get going. They are trapped in an evolutionary dead end they can never break out of.

But many of these animals and plants are found in different habitats and live alongside other species that *have* continued to evolve. Sharks live in all the oceans of the world alongside a host of fishy creatures that have evolved way beyond them, and horsetails grow alongside other much more advanced plants. To say chance mutation has never thrown up genetic improvements for these species begs the question of why it obliged for most others.

And there's no question that the conditions in which some of these living fossil species exist have changed dramatically during their existence. Fossil dragonflies from 325 million years ago look exactly the same as today's. Dragonflies are considered to be among the first, if not *the* first, insects – indeed, the first creatures – to develop flight. And they have carried on happily unchanged, seeing the rise and fall of the dinosaurs 230 to 65 million years ago, the appearance of mammals 190 million years ago and birds 150 million years ago.

Today's dragonflies have to survive against predators, chiefly birds and web-spinning spiders, but as the first creatures to take to the air, they didn't have to contend with them originally. There were simply no birds, flying dinosaurs or mammals. Spiders with the ability to spin suspended webs to catch flying prey only appeared 200 million years ago. But dragonflies nevertheless survived throughout that time, and the appearance of those predators, *without ever adapting*. In other words, dragonflies 325 million years ago were fully adapted to life in the twenty-first century. This flatly contradicts the conventional notion of evolution being an 'arms race' between predators and prey.

All these examples demonstrate that evolutionary theory is so flexible that it, too, has the ability to adapt itself to any given situation. If two species in similar environments are different, that's divergent evolution; if they are the same, that's convergent evolution; if a species hasn't changed at all, that's stasis. It's all OK. It all fits. Actually it doesn't, but it will have to. Nothing is as evolutionarily complacent as evolutionary theory itself.

There are more examples of this reasoning. Many species have adapted so specifically to a particular habitat that they can survive there and there only. The evolutionary explanation is that the species has carved out its own unique niche and it alone is capable of exploiting it. This means that the species has no competition, and so it thrives. On the other hand, there are animal and plant species living in a wide variety of environments. In these cases, we're told, evolution has favoured flexibility because that increases the chances of survival, as adaptation that is too specific puts all the species' eggs in one basket.

So which is it to be: evolution tending towards increasing specialization or greater versatility? Naturally, the standard answer is that different things work for different species, so each case has to be judged on its own merits. It's here that we begin to see the infamous circularity at work. Survivors survive. The eminent philosopher of science, Karl Popper, noted scathingly (his emphasis):

Take 'adaptation.' At first sight natural selection appears to explain it, and in a way it does; but hardly in a scientific way. To say that a species now living is adapted to its environment is, in fact, almost a tautology . . . Adaptation or fitness is *defined* by modern evolutionists as survival value, and can be measured by actual success in survival: there is hardly any possibility of testing a theory as feeble as this.[45]

This sloppy reasoning also prompted Popper to say (his italics), 'I have come to the conclusion that Darwinism is not a testable scientific theory, but a *metaphysical research programme* – a possible framework for testable scientific theories'.[46] He argued that Darwinism became universally accepted because:

Its theory of adaptation was the first nontheistic one that was convincing; and theism was worse than an open admission of failure, for it created the impression that an ultimate explanation had been reached.

Now to the degree that Darwinism creates the same impression, it is not so very much better than the theistic view of adaptation; it is therefore important to show that Darwinism is not a scientific theory, but metaphysical.[47]

Labelling natural selection metaphysical is, of course, an exquisite irony.

Even John Maynard Smith, a self-confessed 'unrepentant neo-Darwinist',[48] declared his own distaste for the 'belief that if some characteristic can be seen as benefiting a species, then all is explained'.[49] But sadly that's all we get from his peers.

Even being generous, the neo-Darwinian theory of evolution is nowhere near as solid as its proselytizers pretend. It has many more gaps and areas of astounding vagueness than they would ever admit to the public. It is, in fact, a startlingly anaemic theory, manifestly failing to explain *any* of the really major events in the development of life on Earth. Its 'explanation' of much of the rest is no more than a description, backed up with circular reasoning that assumes the correctness of the theory in the first place. This is analogous to physicists claiming to have a theory of everything that was absolutely complete – except for its failure to explain gravity or the behaviour of subatomic particles.

Evolutionary biology is, surely, unique among the sciences in that it uses gaps in its knowledge to *support* its fundamental theory, arguing that, since nobody can prove it wrong, the theory must stand. It is less a theory than a default position.

Undeniably, molecular biology has made huge strides in understanding what makes living things tick, particularly the workings of DNA and genes. Although a multitude of mysteries still remain unsolved, the essential laws of genetics – how genes determine the form of an organism and govern its survival, and its role in heredity – have been thoroughly tested scientifically.

What has not been proven, and it is hard to see how it ever could be, is the proposition that random mutations in genes, and random mutations alone, drive evolution. Ever since Darwin, the basic argument has been that chance is responsible for evolutionary change because it *must* be, since self-evidently no non-chance factors can possibly exist. If some other factor was involved it would have to be, by definition, supernatural and everything must be explicable in mechanistic terms? There can be no suggestion of purpose, let alone design. From his ivory tower of certainty, Richard Dawkins writes in the final chapter of *The Blind Watchmaker* (1986) (his emphasis):

My argument [in this chapter] will be that Darwinism is the only known theory that is in principle *capable* of explaining certain aspects of life. If I am right it means that, even if there were no actual evidence in favour of the Darwinian theory (there is, of course) we should still be justified in preferring it over all rival theories.[50]

The reason for this public show of certainty and the unwillingness to admit that there are gaps in current knowledge are understandable – at least to some extent. Evolutionary biologists would probably be more candid about the weak spots in their discipline if it was not for the fact that vested religious interests are ready to pounce at any sign of wavering.

One consequence of this approach is that evolutionary theory has become the bedrock of scientism – science as an ideology as opposed to a method of investigating the world. As Simon Conway Morris observes, 'More than one commentator has noted that ultra-Darwinism has pretensions to a secular religion.'[51] Backsliding and expressing honest doubts about the completeness of the theory is simply not tolerated in biology in the same way it is, for example, in physics.

There is no question, however, that the biblical model should be rejected. God did not make all species complete as they are today in a week. Evolution, in its widest sense, is an established fact, even though many of the details about the precise mechanisms and forces that drive it remain highly debatable. The evidence for natural selection alone deals a deathblow to creationism, although mutation does potentially allow God to slip back in to decide what changes will (or might) work so that tweaks can then be made to DNA. But it is surely something of a demotion, and rather demeaning to an allegedly all-powerful deity. Why should the God of Judeo-Christian tradition be restricted to working in this way?

Rejecting the God of Genesis does not preclude some form of 'soft' design, an active but *limited* creative force at work. In fact, certain aspects of evolutionary history would be easier to explain if such a force existed. Everything considered, neo-Darwinism is neither the coup de grâce to all design theories, nor the atheist epiphany it is supposed to be.

According to Dawkins, once you properly understand neo-Darwinian theory, you *know* there is neither God nor any kind of supernatural force at work in the universe. However, the man who originally formulated the neo-Darwinian synthesis – of which Dawkins is the eager acolyte – saw it very differently. In fact, this largely unacknowledged genius would have had no problem with the thrust of this part of the book . . .

THE GOD GIVER

It comes as something of a shock to discover that Darwin's ideas were far from the overnight success most people believe them to be. As science is no exception to the rule that history is written by the victors, today we have the impression that the publication of Darwin's *On the Origin of Species* changed everything at a stroke. In fact it took almost a century for his ideas to become the scientific givens that they are now. Until as late as the mid-1930s *most* biologists and palaeontologists considered that Darwin was, at best, half right, and factors other than natural selection played a part in evolution. Although a number of influential biologists quickly embraced natural selection, many either rejected it or regarded it as an interesting but unproven hypothesis. Palaeontologists in particular refused to accept Darwin's theory because it failed to fit the fossil record. [52]

A great irony is that the rise of genetics in the first decades of the twentieth century was originally thought to demolish Darwinism. The whole basis of genetics was that genes are fixed and unchangeable units of heredity – the biological equivalent of atoms – while Darwinism required them to vary. The neo-Darwinian synthesis was the result of reconciling genetics with Darwinism, laying the foundation for everything that has come after. It was the recognition that genetic mutation was the cause of the small, individual variations that natural selection seized on and honed.

Ernst Mayr and science historian William B. Provine sum up the rapidity of the change in attitude in their introduction to *The Evolutionary Synthesis* (1980):

> In the early 1930s, despite all that had been learned in the preceding seventy years, the level of disagreement among the different camps of biology seemed almost as great as in Darwin's day. And yet, within the short span of twelve years (1936–47), the disagreements were almost suddenly cleared away and a seemingly new theory of evolution was synthesized from the valid components of the previously feuding theories.[53]

The momentum has carried on ever since. But just what happened over those dozen years, and why did Darwinism come out on top after nearly a century in the wilderness?

The surprising difficulty in answering this question is shown by the number of conferences called to discuss the events of those years. *The Evolutionary Synthesis* was a collection of the papers delivered at one such event organized by the American Academy of Arts and Sciences in 1974. A similar gathering was held in 1981 at Bad Homburg in Germany to discuss the rather syntactically tortured question: 'How complete and how stable is, and has been, the evolutionary synthesis, or "neo-Darwinism"?'

It was there that Stephen Jay Gould delivered his paper on 'The Hardening of the Modern Synthesis', which covered the crucial 1936–47 period. After surveying the process of theory-hardening, he came to the more important, but problematic, question of *why* it had happened in the first place, admitting: 'I now arrive at the point where I should give a conclusive and erudite explanation of why the synthesis hardened. Yet truly, I do not know.'[54]

He offered two possible explanations. The first he called the 'heroic' version, which is that evolutionary biologists came up with the right answers through an objective evaluation of the evidence. The second, the 'cynical' version, is that the advocates of natural selection were themselves guilty of selection, by picking only the evidence that fitted the emerging consensus and dismissing the rest:

> Since the world is so full of a number of things, cases of both adaptation and nonadaptation abound, and enough examples exist for an impressive catalogue of partisans of either viewpoint. In this light, historical trends in a science might reflect little more than mutual reinforcement based on flimsy foundations.[55]

If the cynical version is right, Gould pointed out, it might be preventing a proper understanding of evolution by ignoring factors other than natural selection. But which is right? Gould concluded once again: 'The only honest answer at the moment is that we do not know.'[56] That was 1981, but the situation is still pretty much the same. Neo-Darwinism still dominates, but perhaps that's because it refuses to look too closely at potentially hostile data.

The 'hardening' of the theory was almost entirely due to one man. Theodosius Dobzhansky (1900–75) was a Russian-born, naturalized American biologist and it was his 1937 landmark book *Genetics and the Origin of Species* that showed the way to reconcile natural selection and genetics.

Flick through the pages of any academic book on neo-Darwinian theory and Dobzhansky is a star, acknowledged for his revolutionary insight that laid the foundation for everything that came after. But look in any more popular account and you'll be lucky to find him so much as mentioned. He doesn't rate a single reference in Dawkins' *The Blind Watchmaker* or *The Greatest Show on Earth* (although he is mentioned in passing in *The Ancestor's Tale*, as 'the great evolutionary geneticist'.)[57] There may be a good reason for the difference between the way specialists talk about him among themselves and the relative silence in their public pronouncements. It's quite simple. Dobzhansky is something of an embarrassment because he was unashamedly a devout Christian. (Neatly, Theodosius means 'God-giver'.)

Not only was he an active member of the Eastern Orthodox Church, but he saw no incompatibility between his faith and his belief in evolution. He even saw evolution as God's way of expressing and achieving his purpose, writing in 1970 that, 'man was and is being created in God's image by means of evolutionary developments'.[58]

Dobzhansky regarded evolution as a 'creative process'.[59] To him this did not compromise the essential blindness of natural selection: chance was an important part of the process. He thought that the putative universal designer – to him the Christian God – had set in motion a system that enabled life to develop and find its own way. He preferred to talk of natural selection as *groping* its way forward, having 'tried out an immense number of possibilities and . . . discovered many wonderful ones. Among which, to date, the most wonderful is man'.[60]

Even this was part of his much wider vision. In the words

of Greek geneticist Costas R. Krimbas, one of Dobzhansky's research students in the late 1950s, he:

> . . . recognized that organic evolution was part of a cosmic process that comprised the birth and evolution of matter and stellar bodies, the appearance and evolution of life, and finally the genesis of humankind. Every time the process passes from one stage of complexity to the next, it transcended itself, first in the transition from matter to life, and then in the genesis of humans, the transition from material life to cultural life.[61]

Dobzhansky took the image of groping forward from Pierre Teilhard de Chardin, the French Jesuit palaeontologist who we mentioned briefly earlier, writing: 'This is a splendid, though somewhat impressionistic, characterization of evolution moulded by natural selection.'[62]

Teilhard de Chardin (1881–1955) was a paradoxical combination of Jesuit priest and evolutionary theorist. His eagerness to combine evolution with Catholicism was not shared by his fellow Jesuits, who wasted no time in posting him to China to prevent him lecturing on the subject. There he was part of the team that discovered Peking Man, fossil remains of *Homo erectus* over half a million years old. Teilhard was forbidden to publish any philosophical works or, on his return to Europe twenty years later, to apply for academic posts. As a result, he went into self-imposed exile in New York. His classic work, *The Phenomenon of Man* (*Le phénomène humain*) was published shortly after his death in 1955, when the ban expired with him.

Teilhard saw the universe as absolutely purposeful, the aim of matter being to engender life and the goal of life being to attain consciousness. He argued that human consciousness would eventually create a planetary spiritual entity that he called the noosphere, which would eventually link with extraterrestrial intelligences; life and mind would then permeate and take control of the universe. The goal of the entire process was the 'Omega point', at which creation reunites with its creator. To Teilhard this meant reunification with the Christian God. He declared that 'evolution is an ascent towards consciousness – therefore it should culminate forwards in some sort of supreme consciousness'.[63]

Although most of his concepts had already been around for thousands of years, Teilhard's contribution was to link them with twentieth-century ideas, particularly those from the biological sciences. The idea that the divine is present in everything and that creation is unfolding and moving determinedly towards a specific end underpins many ancient mystical systems – ironically for Catholic Teilhard, most of them Gnostic. It also very much underpins Hermeticism – especially the prime role of mind in the evolution of the cosmos – and the all-important arcane school of Heliopolis from which it developed.

Tantalizingly, there is even a specific connection, albeit an indirect one, between Teilhard's ideas and the great Egyptian school. The same underground stream sweeps certain luminaries along throughout the millennia. Teilhard's formative influence was the philosophy of Henri Bergson, particularly his *Creative Evolution* (*L'Évolution Créatrice*), which Teilhard read just before his ordination in 1912. Bergson (1859–1941), in turn, was heavily influenced by the works of Plotinus,[64] the 'Neoplatonic' philosopher who we argue was more accurately neo-Egyptian given that he ultimately drew his inspiration from the religion of Heliopolis. Bergson also gave a series of lectures on the 'numerous and impressive' parallels between Plotinus' system and Leibniz' theory of monads.[65]

Teilhard's ideas on purposeful evolution were surprisingly

influential, particularly in the French-speaking world, and remain cautiously debated by scientists such as Christian de Duve, John Barrow and Frank Tipler. The latter two wrote in *The Anthropic Cosmological Principle* that 'the basic framework of his theory is really the only framework wherein the evolving cosmos of modern science can be combined with an ultimate meaningfulness to reality.'[66]

Teilhard de Chardin obviously represents the polar opposite to Richard Dawkins, which is deeply ironic given that Dobzhansky, founder of Dawkins' discipline, embraced Teilhard's creative evolution. Not only did Dobzhansky greatly respect Teilhard's philosophy, in the 1960s he even became President of the American Teilhard de Chardin Society. Significantly, however, he did not begin as a 'Teilhardist' and tailor his work in evolutionary biology to fit. Quite the reverse. It was his work on the neo-Darwinian synthesis – especially the implications of a creative element in evolution – which led him to Teilhard. To Dobzhansky the genetic system was fully compatible both with the idea of a creative, intelligent universal power *and* a universe evolving towards an ultimate goal.

However, the mysteries discussed in this chapter suggest that even this fails to present the complete picture. As Dobzhansky saw it, God made DNA and left it to get on by itself, confident it would eventually reach its destination. But perhaps Dobzhansky stopped short of a full answer. It does appear that other events, elements of 'luck' with no connection to the genetic system, were contrived to get life past particular blocks on the evolutionary road . . . perhaps with GUD's helping hand.

The belief that a purely mechanistic explanation must lie behind the processes that shape evolution might hold up if the sciences generally had found no evidence of design in the rest of creation. But they have. Physics, in particular, has moved on since the mechanistic Victorian science in which Darwin advanced his theory. Biology hasn't.

To us, towering above all the other tantalizing hints about *true* intelligent design is the uncanny suitability of DNA and its mysterious origins. There does seem to be something scarily made to order about it. It is not just that a molecule with all the right, miraculous properties for life should have come into being. Whatever process produced DNA did not necessarily have to make something that was also able to adapt to changing conditions. LUCA might have turned out to be an organism that could happily survive and thrive in the conditions of a four-billion-year old Earth, but would die off as soon as those conditions changed.

Similarly, the single-celled life forms that developed from LUCA and populated the planet for the first two or three billion years had limited potential for evolution. Something else had to enter the equation in order to create the revolutionary new type of nucleated cell that enabled more complex organisms to evolve. The standard theory can only ascribe this to sheer fluke. Another fluke started sexual reproduction, speeding up evolution and allowing even more complex forms of life to develop. But sex, too, faced an obstacle that would have limited the genetic diversity that it otherwise allowed had that obstacle not been removed by the appearance of the genes for ageing and, ultimately, death. Is it just us, or does that seem rather contrived?

Such 'luck' suggests that a proper understanding of evolution *does* require some ongoing creative factor, something somehow capable of comprehending the bigger picture. This, of course, fits elegantly into the designer universe scenario, and supports the evidence from cosmology that the universe was fine-tuned for intelligent life. It also implies, however, that evolution is working towards a specific end, and that the development of ever-more complex life forms is at the core of that process. This in turn implies that humanity represents its cutting edge.

But is there any evidence that human faculties such as intelligence and consciousness are more than just freak products of a blind universe? And could they be in some way actually fundamental to the cosmos?

CHAPTER TWELVE

MIND MATTERS

Despite the bravado and bluster of atheist proselytizers, the very least that can be said is that evolutionary theory is by no means the final proof of their dogma. But there are also certain biological phenomena that apparently provide real evidence for a creative force at work. And this takes modern science ever closer to the core belief and central message of the Hermetica: that every human being is potentially a god, that the universe is alive and that we are all part of its divine long-term plan and destiny.

It seems self-evident when we look at life on Earth and the evolutionary history that science has reconstructed, that life developed from the simple to the complex: from bacteria, through multi-celled micro-organisms, flatworms and insects to mammals, specifically humans. The impression is one of irrepressible progress, creatures becoming gradually more complicated and more able to interact with and modify their environment, besides becoming increasingly intelligent. From this perspective, the human being is the pinnacle of evolution on Earth, the 'most wonderful' result of natural selection to date, as Dobzhansky commented. Karl Popper noted that the ostensibly accidental mutations that drive evolution also uncannily push a species forward – never a step back. Species seem to change by 'sequences of evolutionary changes in the same "direction"'.[1]

Ultra-Darwinists such as Dawkins reject this evolutionary directionality as simple 'species-ism'. We humans think we're the best evolution has produced because we would, wouldn't we, being us. We imagine evolution has progressively produced more impressive species until it made us, its best work to date. But Dawkins argues this attitude is a mistake, if a forgivable one. Objectively, a bacterium or jellyfish is just as 'perfect' a piece of evolutionary design as Professor Dawkins himself (if far more silent). To him, the whole notion of 'higher' and 'lower' forms of life extrapolates too much from simple classifications. And human-like intelligence is by no means an inevitability; the planet got by without it for long enough, after all.

The most that hard-line evolutionists will admit is that natural selection moves a species towards ever more suitable adaptation to its specific environment, but that's not the same as achieving progress for life on Earth in general. Some evolutionists – Dobzhansky being the prime example – do accept the idea of directionality. To them there is no doubt that evolution does tend to produce increasingly complex and more self-aware creatures. Human beings *are* the pinnacle of evolution (so far), although undeniably there is still considerable room for improvement.

PIOUS ATHEISTS AND METAPHYSICAL EVOLUTION

Another scientific champion of directionality who believes that it reveals that there are evolutionary rules and principles yet to be recognized, is Simon Conway Morris, Professor of Evolutionary Palaeontology at Cambridge University's Department of Earth Science, and Fellow of the Royal Society. Despite being a committed Anglican, Conway Morris is equally critical of intelligent design and 'ultra-Darwinists' such as Dawkins, who he describes as 'arguably England's most pious atheist',[2] and being 'angry with God'.[3] As for his own position, Conway Morris sums it up in these words:

Evolution is true, it happens, it is the way the world is, and we too are one of its products. This does not mean that evolution does not have metaphysical implications; I remain convinced this is the case. To deny, however, the reality of evolution and more seriously to distort, deliberately, the scientific evidence in support of fundamentalist tenets is inadmissible.[4]

Conway Morris' special area of interest, convergent evolution – 'the recurrent tendency of a biological organization to arrive at the same "solution" to a particular "need"'[5] – has led him to the conclusion that it happens far more than neo-Darwinian theory dictates.

Conventional evolutionists believe that if we could restart life on Earth, because the evolutionary paths it took were shaped by random factors, the outcome would be very different. In this scenario animals and plants that are nothing like those we are familiar with would populate the world. Human-like creatures may not exist, since nothing would be inevitable. Conway Morris disagrees, arguing that convergent evolution shows that the number of evolutionary pathways is limited and that therefore outcomes are largely predetermined. As he said in a 2007 lecture:

In fact, evolution shows an eerie predictability, leading to the direct contradiction of the currently accepted wisdom that insists on evolution being governed by the contingencies of circumstance.[6]

There are vast numbers of 'macro' examples of convergent evolution, such as the many in Australia discussed earlier, where the similarity between two species is immediately apparent from their appearance. However, Conway Morris demonstrates that many more similarities are not so obvious, since they relate to individual features, the anatomy and workings of a particular organ, say, or even an internal biochemical process. Backed up by a landslide of examples, it is clear that convergent evolution is, if anything, the norm. Evolution plays the same themes over and over again.

Conway Morris takes the prime example of two creatures that are to all intents and purposes as different as two creatures could be: the human being and the octopus (or more generally, mammals and cephalopods, which also include squid and cuttlefish). He observes the two species are so different that octopuses were frequently used in early science fiction – such as H.G. Wells's *The War of the Worlds* – as the model for aliens. Humans and octopuses are the product of two entirely different evolutionary lineages. One a vertebrate, the other an invertebrate, they reflect one of the most fundamental and ancient branchings of the evolutionary tree. It is a very, very long time since we shared a common ancestor; cephalopods are, in fact, a class of mollusc, closely related to mussels and slugs. As we live in an entirely different environment and have gone through completely separate adaptations, it is hardly surprising we should look so different from octopuses. First impressions do seem to confirm conventional wisdom: the further back in time two species shared a common ancestor, the more different they will be now.

The underlying reality is very different. There is startlingly more convergence than one might think. Most obviously, cephalopods have evolved eyes that work, like those of mammals, on the camera principle, with precisely analogous structures performing the same functions. But there are other, equally astonishing similarities. The blood and circulatory system – especially the aorta – of cephalopods is very much the same as that belonging to mammals, and nothing like that of other molluscs. The most intelligent of the invertebrates, the octopus, has evolved a completely different type of brain from mammals', but parts of it are precisely analogous to the mammalian hippocampus and cerebellum. Even the male octopus' sex organ, although positioned at the tip of a tentacle, is structurally very like the mammalian penis, and bears no comparison with other molluscs'. So despite the overall differences in anatomy and its very separate evolutionary path, the octopus is far more 'mammalian' than we imagine.

There are many other examples which demonstrate that convergence is too widespread to be pure chance. The camera eye evolved no fewer than seven times, quite independently. The compound eye of insects has evolved at least four times. Trichromatic colour vision has evolved separately many times, as is the case with New World monkeys and the Australian marsupial honey possums.

That's the big stuff. Convergent evolution also happens to molecules. The biochemical processes that sustain organisms are often complex, and yet distantly-related species have independently developed exactly the same systems. One of the most striking examples involves photosynthesis in plants, which uses sunlight to transform carbon dioxide into oxygen. It's not just important for plants, of course: as Conway Morris points out, photosynthesis literally underpins the whole biosphere. Most plants use a chemical process known as C_3 photosynthesis, but many use an alternative, the much more complex C_4. This is an adaptation to an environmental change: over the last ten million years or so there has been a dramatic drop in carbon dioxide in the atmosphere, making life difficult for many plants. But not all the plant species that use C_4 photosynthesis have, as we might expect, evolved from the first one to hit on it. Despite its complexity, this system has evolved quite independently *at least 31 times*.[7]

Conway Morris' *Life's Solution* (2003) is packed with the most extraordinary examples of convergent evolution among animals, insects, plants and bacteria. He argues that the unexpected prevalence of convergence shows that, rather than evolution picking paths from a limitless number of possibilities, it continually finds and follows the same well-worn grooves. To him the evidence overwhelmingly suggests that the phenomenon reveals the existence of some factor that neo-Darwinism has yet to recognize. Conway Morris concludes by saying: 'It seems to me that evolution very much has directionalities, and in that sense it has destinations.'[8]

Restricted options imply that certain biological phenomena will *inevitably* evolve. Re-running the history of life on Earth would end up with creatures and plants pretty similar to modern ones. Conway Morris argues one of the outcomes is not only intelligence but that 'the constraints of evolution and ubiquity of convergence make the emergence of something like ourselves a near-inevitability'.[9] According to the conventional view, humanity is just an insignificant accident, lucky to be here. But Conway Morris makes humans the focus of the universe, the very reason it exists. If evolution was always a journey towards humankind, then we are very special indeed.

'SMALL BUT NOT STUPID'

Further evidence that the evolutionary trajectory is aiming for creatures like ourselves comes from recent discoveries about intelligence and human-like behaviour throughout the animal kingdom. These revelations are finally overcoming humanity's belief that our species is set apart from the rest of nature, the only creature able to properly think and feel. Now we know that intelligence – the ability to solve problems and react creatively to changing circumstances – is widespread in nature. We may or may not be alone in the universe, but we are not alone on our home planet.

Swiss anthropologist Jeremy Narby's book *Intelligence in Nature* (2005) relates how intelligent, problem-solving behaviour is not just found in higher animals such as primates and birds, but even among butterflies and such lowly life forms as slime moulds, which can negotiate mazes to find food. Recognizable intelligence is a feature of even the most primitive organisms. Amoebae engage in co-ordinated, cooperative behaviour to hunt their prey in packs. In 2007 James A. Shapiro, a bacterial geneticist at the University of Chicago, wrote a landmark paper entitled 'Bacteria are Small but not Stupid', a plea for the recognition that bacteria are sentient beings because they 'continually monitor their external and internal environments and compute functional outputs based on information provided by their sensory apparatus'.[10]

Other research, including that of Jonathan Balcombe, as set out in his book *Second Nature* (2010), has shown that not just intelligence, but other characteristics we usually think of as exclusively human, such as awareness of death, a sense of grief, even a sense of fun, are an intrinsic part of animal lives. Although elephants' capacity to grieve over the death of a herd member is well known, recently they have been recorded as using ritual – such as passing around sticks – at the death of a loved one. Chimpanzees in a zoo, meanwhile, have been observed to stand silently in a circle and cry as a deceased friend was carried past. As Balcombe repeatedly emphasizes, animals are not just living, *they have lives.* And the complex and often touching nature of their lives reveals their innate intelligence and an awareness of more than simply the mundane and the present.

Many evolutionary biologists, such as Ernst Mayr, believe

that while life might be common in the universe, *intelligent* life is so improbable it is virtually exclusive to Earth. Others, such as Christian de Duve take issue:

> Conscious thought belongs to the cosmological picture, not as some freak epiphenomenon peculiar to our own biosphere, but as a fundamental manifestation of matter. Thought is generated and supported by life, which is itself generated and supported by the rest of the cosmos.[11]

Simon Conway Morris is of the same mind. In 2007 he gave the annual Gifford Lectures at the University of Edinburgh – the series was instigated in 1887 by Lord Gifford to explore the theological implications of scientific advances – under the banner title of 'Darwin's Compass: How Evolution Discovers the Song of Creation'. The second of the six lectures was tellingly subtitled 'The Inevitable Evolution of Intelligence', while in another he declared the emergence of life, and of intelligence and of intelligent beings 'really is set into the entire fabric of the cosmos'.[12]

The universality of intelligence and other mental and emotional phenomena once thought to be exclusively human supports the idea that nature – and indeed the universe – *wants* to produce self-aware organisms able to take control of their environment. But even given our ever-closer kinship with other species, there does seem a real gulf – a quantum leap – that separates us from even our nearest evolutionary relatives. We wear clothes, tell stories, glory in language, explore our own planet and even deep space with increasingly sophisticated science.

The Dawkins school of thought doesn't deny that we humans are in a unique position, especially when controlling our evolutionary destiny, but contend that it is all just an accident, and there's nothing inherently special about our abilities. Others disagree. Michael Polanyi, Hungarian philosopher of science declares:

> It is the height of intellectual perversion to renounce, in the name of scientific objectivity, our position as the highest form of life on earth, and our own advent by a process of evolution as the most important problem of evolution.[13]

And Simon Conway Morris once again defends human greatness:

> . . . incipient 'human-ness' is clearly visible in a wide variety of animals, be it expressed in terms of tool-making, singing or even awareness of death. Yet in no case has it 'crystallized'. We stand alone, feet on the ground, head towards the stars.[14]

But is it intelligence that the universe seems compelled to seek – or is it consciousness? At its most basic, intelligence is the ability to adapt behaviour in response to data received by the senses, the type of intelligence exhibited by bacteria and slime moulds. That kind of problem-solving intelligence doesn't necessarily require self-awareness or the ability to reflect. Slime moulds can learn to negotiate mazes, but still demonstrate nothing like human consciousness. Slime mould philosophers are very rare – as far as we know.

If the universe is designed for life then there must be a reason – something that life is needed for. Cosmic evolutionary theorists such as Teilhard de Chardin argue that consciousness is what life – and even matter – is ultimately striving for. Carl Sagan famously declared that 'we are a way for the Cosmos to know itself'.[15] Are we really? Does the universe for some reason *need* conscious entities? And if so, why?

Very bizarrely, there is real scientific evidence that the purpose of the universe is indeed to produce conscious, thinking entities – for a very good reason. It needs us to bring the universe itself into being . . .

We are now entering a very strange world indeed.

GLOBAL EXCITEMENT

We know what we mean by 'consciousness' because we all have it and never stop using it until the day we die – and perhaps not even then. But can this elusive invisible thing that shapes our personalities and all of our utterances be defined and explained scientifically? Where does it reside, how does it work, and how does it relate to the world around us? Unlike DNA, which creates and maintains our bodies, it is impossible to locate or analyse consciousness under a microscope, although it is assumed to be connected with the brain.

Since the late 1980s there have been many attempts to explain consciousness in terms of quantum processes. One of the first was by Oxford University mathematician Roger Penrose – author of *The Emperor's New Mind* (1989) – who went on to collaborate with Stephen Hawking. Penrose said: 'There is a certain sense in which I would say the universe has a purpose. It's not there just somehow by chance.'[16]

However, most attempts to link consciousness and quantum theory tend to be fuzzy and speculative, which is not totally surprising as they seek to explain one nebulous issue in terms of another. Basically, although there is a groundswell of feeling that consciousness will prove to be explicable in terms of quantum processes rather than as a chemical product of the brain or similar phenomenon in the 'macro' world, it is still very early days. But if the quantum route does turn out to be fruitful, the implications are enormous. It will mean human consciousness is intimately connected with the physical world at a very fundamental level, an astonishing – even apparently magical – scenario, with which the old Hermeticists would be totally at home. And this fits with accumulating evidence from the physical sciences that the very existence of consciousness can and does have a tangible, measurable effect on the world of matter.

One of the physicists drawn into the study of consciousness was Dick J. Bierman of the University of Amsterdam. From physics he moved into artificial intelligence, which naturally involved a study of cognition – how the mind picks up and processes information about the external world. This led him into the study of consciousness and its relationship with quantum physics. In fact, he got drawn even further into the physicists' forbidden realm of parapsychology, the study of alleged weird abilities and events, known collectively as *psi*. He reasoned that psychic abilities could be a possible manifestation of the interface between consciousness and the quantum world.

But was Bierman brave or just foolhardy to enter the world of parapsychology? Even the word itself is a turn-off to self-confessed rationalists. Ever since attempts began to scientifically test claims of psychic abilities – telepathy, precognition and psychokinesis, or mind over matter – the scientific world has opposed not just the claims, but even the idea of testing them (unless the tests disprove the claims, of course). But why the prejudice?

The fundamental objection is that such phenomena just *can't* exist since they violate the most basic, common sense principles that underpin our understanding of the material world. Telepathy upsets the rule that there must be a physical link between two objects for them to transmit information to each other. Precognition stands the concept of cause and effect on its head. Psychokinesis, or the alleged effect of mind over matter, is the ultimate horror, since it violates pretty much all the basic principles, including the laws of energy conservation. If real, psychokinesis would mean that it is possible to conjure energy out of nowhere. Unsurprisingly the scientific community at large has a problem with the paranormal. Such things can't possibly *be*.

However, these rules only apply to the macro world of the atom and above. As understanding of the subatomic, quantum world has grown over the last century, it has become increasingly obvious that the common-sense principles with which we judge the world have no jurisdiction down there. There, effects sometimes precede causes ('backward causation'). Particles can jump from one energy state to another without apparently getting the energy from anywhere. Experiments have shown that two particles created by the same event – a collision in a particle accelerator, for example – remain in some weird way connected, continuing to influence each other even when far apart and no longer linked in any way. And they can do so instantaneously, even seeming to breach the ultimate barrier of the speed of light.

Of all these violations of common sense, the most relevant to this discussion are the ones that relate to time. It may seem odd to most of us, but the fact that time usually flows in just one direction is a real puzzle to physicists, since there is no discernable reason for this according to the laws of physics. In theory many physical processes should be able to work in either direction. Whole conferences have been devoted to the problem of 'time asymmetry', such as one organized – somewhat unexpectedly – by NATO in 1991 in Magazan, Spain where celebrity scientists such as Stephen Hawking and John Archibald Wheeler delivered papers.[17]

In his 1988 paper, 'A World with Retroactive Causation', Bierman argues that even in the macro world, 'there is empirical evidence that effects can precede causes'.[18]

308

He argues that no paradox is involved, and that his findings fit the discoveries of quantum physics. Describing its implications 'far-reaching' is something of an understatement.

Given that subatomic particles have been demonstrated to act fast and loose with supposedly inviolable physical principles, it seems almost unsophisticated to insist that they have to be obeyed everywhere else – with no exceptions. The ever-perceptive Paul Davies makes the point that whereas scientists are quite happy to explore ideas of backward causation and instantaneous communication between unconnected particles, 'it is only when the end state involves life and mind that most scientists take fright and bale out'.[19] In other words, it is fine for a subatomic particle to 'see' into the future, but not for a human being.

A handful of physicists – most prominently British professor Brian Josephson, joint winner of the Nobel Prize for Physics in 1973 for his work on superconductivity – has openly accepted the reality of psi and is actively seeking a quantum explanation. As a result he is now head of the Mind-Matter Unification Project at Cambridge's Cavendish Laboratory. Josephson is fond of using the Royal Society's motto, *nullius in verba* – our favourite translation being 'take nobody's word for it' – against scientists who dismiss parapsychology without deigning to look at the evidence. In an interview for *New Scientist* in 2006 on this very topic he railed:

I call it 'pathological disbelief'. The statement 'even if it were true I wouldn't believe it' seems to sum up this attitude. People have this idea that when something can't be reproduced every time, it isn't a real phenomenon. It is like a religious creed where you have to conform to the 'correct' position.[20]

He added: 'These things are not hard to prove, they're just hard to get accepted.'[21]

The general trend towards linking consciousness and quantum physics promises parapsychologists real hope. If mind and matter prove to be connected at that deep level it could offer an explanation for psi that keeps it within physical laws. This is the line taken, for example, by leading American parapsychologist Dean Radin in *Entangled Minds: Extrasensory Experience in a Quantum Reality* (2006).

The most exciting discoveries to emerge from parapsychology in recent years do appear to confirm a link between consciousness and the material world at the quantum level. This began serendipitously during research by Bierman.

In the mid-1970s Bierman pioneered the use of Random Event Generators (REGs, also called Random Number Generators) in psi experiments. The advantage of REGs is that they circumvent one of the main problems in evaluating psi experiments. To substantiate claims of extraordinary abilities, the outcome of an experiment has to be compared to chance, which is why all too often parapsychology disappears into a fog of tedious calculations and statistics that become hard to interpret – or have several possible interpretations. Bierman first used a REG in experiments where volunteers tried to mentally influence the output. It was therefore easy to see whether the output had deviated statistically from chance – as indeed it had, unequivocally.[22]

In 1995 Bierman was using an REG in a house in Amsterdam where poltergeist activity was allegedly taking place, testing whether the REG behaved differently when the invisible hooligan was at work. When the results were analysed for one particular day they did indeed show a ninety-minute period of non-random output – but puzzlingly this related to no spooky goings on in the house.

Bierman and his team realized it coincided with something rather more mundane: the 1995 UEFA Champions League final, in which Ajax – the famous Amsterdam football team – was playing AC Milan. Even more tantalizing, the moment of greatest non-randomness coincided with Ajax scoring the only goal of the game.[23]

The REG output was obviously affected by some aspect of the game, the most obvious candidate being the country's intense focus and collective excitement. The same effect has been found since, for example in a 2004 study by German researchers at the Institut für Psycho-Physik in Cologne, during an important football match in the city.[24] This suggested a completely new avenue for research, not involving the special mental states associated with psi but the collective workings of ordinary people's consciousness in everyday situations.

Bierman's accidental discovery particularly excited a group of American parapsychologists, including Dean Radin. Seeking the same effect in 1996, he and his colleagues began the REG monitoring of mass events such as the Oscars, the Superbowl and the opening and closing ceremonies of the Atlanta Olympics – anything with television audiences of many millions. Although the results were variable, they seemed to confirm Bierman's discovery. This encouraged them to follow a new line. Rather than picking selected events in advance, they decided to set up a system to permanently monitor fluctuations in global randomness. This way they could find out if a similar effect coincided with unplanned news events – major disasters or the death of an international celebrity, for example.

The idea was given a dry run with the television coverage of Princess Diana's funeral in August 1997, which obviously had the advantage of being both global and live. Using twelve REGs, they found deviations of 100 to 1 against chance in their output. Cannily, they used Mother Teresa's funeral a few days later as control. This was also broadcast live, but the peaceful death of an old lady, however much respected, carried little of the raw emotion associated with the demise of a glamorous young princess and mother in horrifying circumstances. This time they found no effect.

Encouraged by these preliminary results, the Global Consciousness Project was created in 1998, funded by the Institute of Noetic Sciences, where Radin is a senior researcher, and headed by Roger Nelson of Princeton University. The Institute of Noetic Sciences is the California-based research institute founded in the 1970s by Apollo astronaut Edgar Mitchell, the sixth man to walk on the moon. ('Noetics' comes from the Greek *nous*, the faculty of 'inner knowing', which has no exact equivalent in English. The word is liberally sprinkled throughout the Hermetic texts.)

There is now a network of some 65 REGs – nicknamed 'eggs'– located all over the world, from large American cities to remote Pacific islands, connected by the Internet. All the REGs do is continually churn out their counts, one per second, day in, day out. The data from each egg is downloaded every five minutes to a server in Princeton, which is accessible to any interested party. The results are then analysed for periods of non-randomness, either from individual or all eggs, which are then compared to world events. Conversely, when major news events occur, the REG data is examined for signs of non-randomness.

One of the most elegant aspects of this set-up is that because the data from all the eggs has to be grouped together, put through a series of statistical analyses and then plotted on graphs before any anomalies can be noticed, it isn't readily apparent just from the streams of numbers that anything interesting has happened. The analysts can't bias the results even by subconscious selection of the data. Dates and times of all the major global events – both

pre-planned such as sporting fixtures and awards ceremonies or random occurrences like major disasters – which
happen within a particular period can be listed from an
independent source such as the annual review of a news
service. The data from the eggs during that period can be
analysed independently, and then the two compared for
correlations. And the calculations can be checked on
request.

The results have been unequivocal. The periods of
anomalous non-random output coincide with times of
major global events. Dean Radin demonstrated this most
vividly in 2001, when the REGs' output deviated from pure
chance many times, but one day above all stood out for the
sheer size of the deviation . . . 11 September, when the eyes
of a horror-struck world were riveted on television footage
of the terrorist attack on New York's Twin Towers and its
sickening traumatic aftermath. Likening the sharp peaks
and troughs on the graph to the ringing of a bell, Radin
wrote that, 'in metaphorical terms, our bell rang more
loudly on this day than any other day in 2001'.[25]

Even more compelling evidence that the REGs were
measuring something real came from a more detailed
analysis showing that it wasn't just the amalgamated data
from all the eggs that 'rang the bells'; all the individual eggs
around the world rang that day. As Radin declared:
'*Something*, perhaps changes in mass attention, caused the
random data to behave in a dramatically non-random way
on 9/11, whereas it behaved normally on other days'.[26]

Inevitably, critics claim that the apparently striking
results of the Global Consciousness Project are due to
methodological flaws in analyzing the data. But given the
sheer amount of accumulated information from the last
decade, it is hard to see the results as demonstrating anything other than a real effect. Human consciousness really
does seem to have a tangible effect on the material world.

So given the enormous implications, why isn't this 'global coherence effect' much more widely known? Probably because to non-scientists its significance might be hard to grasp and even seem rather dull. After all, this is not exactly moving mountains by the power of mind alone. The experiments show that the focused attention of millions of people is needed to cause just tiny fluctuations in a few REGs – which is not even in the same league as one dramatic spoon-bending.

What exactly do these results tell us? The Global Consciousness Project team use them to support the idea of the evolution of a planetary consciousness – the noosphere, a term borrowed from Teilhard de Chardin. However, that may be extrapolating way too much from the current data. It is true that such an effect is exactly what Teilhard and others would have predicted, and it may indeed turn out to be a sign of the emergence of a global consciousness. But right now the evidence simply doesn't stretch that far.

What can be said at the moment is that the network of REGs is not being *deliberately* influenced by the massed minds of the people on the planet, only a relative handful of whom even know it exists. The REGs can only be registering a side-effect of something else, something that people are unaware of doing. And the effect can't be confined to the REGs; if their output is less random, the effect can only be because all and any random processes are being smoothed out in some way. When a large number of people pay attention to the same thing, for some as yet unknown reason the world becomes more ordered, particularly at the quantum level where randomness and unpredictability rule. It is not even deliberate; it just seems to be the effect that consciousness creates, simply by existing.

Perhaps what is even weirder is that this is also the thinking of certain top physicists, who propose that consciousness – human or otherwise – is literally what keeps the universe in place. And even that consciousness created the universe in the first place.

'THE MYSTERY WHICH CANNOT GO AWAY'

We all know the world of quantum mechanics is head-spinningly weird, but it does have a clear relevance to our understanding of life, the universe, everything – and humanity's role in all of it. And despite the implications of quantum theory being so left-field that even Einstein had problems with it, it does provide some potential clues in our search for the mind of God – or, indeed, our Great Universal Designer, GUD.

Einstein clashed, albeit in a friendly fashion, with Neils Bohr, the great champion of quantum theory, in a debate that went on for nearly thirty years. John Archibald Wheeler, who studied under both luminaries, wrote in his autobiography:

> These two giants, full of admiration for each other, never came to agreement. Einstein refused to believe that quantum mechanics provides an acceptable view of reality, yet he could never find an inconsistency in the theory. Bohr defended the theory, yet he could never escape being troubled by its strangeness. Reportedly, once when Einstein remarked, as he liked to do, that he could not believe that God played dice, Bohr said, 'Einstein, stop telling God what to do'.[27]

One of the most bizarre aspects of quantum mechanics is that it recognizes an intimate relationship between the mind of an observer and what happens at the quantum level. It is really just a question of how deep the relationship goes.

The classic example comes from the famous 'wave-particle duality' conundrum, the recognition that subatomic particles (in most experiments photons, the particles of light, but it applies to all of them) sometimes behave like particles and sometimes like waves. Richard Feynman called the enigma 'the mystery which cannot go away'.[28]

The classic demonstration of wave-particle duality is the renowned 'double slit' experiment, the earliest version being carried out as long ago as 1803, by the woefully little-known English polymath Thomas Young (1773–1829). The scientist, physician, philologist and Egyptologist disproved the prevailing view, established by Newton, that light was made up of particles, by demonstrating it was really a wave. By shining a single beam of light through two narrow slits onto a screen, Young showed that bands of light and dark appeared. Such interference patterns are only explicable if light moves in waves: the light passes through both slits and, just like water in similar circumstances, the two waves emerging from each slit either cancel each other out or reinforce each other to produce the interference pattern.

However, when quantum theory came along a century later, physicists realized that light ought to be made up of particles after all. Young's interference patterns were not initially too much of a problem, since photons en masse could work in waves, just as sand can be made to ripple in a wave-like fashion. The real difficulties began when even just a single photon at a time was fired at the screen and the same interference patterns built up.

The results were totally counter-intuitive. If one slit is closed and a beam of light shone through the other then – as expected – just a single sharp line appears on the screen. If the slit is closed and the other opened, then again a single line appears in a different place on the screen. But if both slits are open at the same time, you get the interference patterns – even when just a single photon is involved. The photon seems to be interfering with itself, so to speak. As Paul Davies comments: 'It's almost as if the photon can be in two places at once, that is pass through both slits.'[29]

It gets odder. The outcome – whether light behaves as a wave or particle – depends on how the photon is detected after passing through the slits. When a light-sensitive screen such as a photographic plate is used, the interference patterns typical of a wave appear. If two telescopes or similar devices are instead trained separately on each slit, then every individual photon will be detected by only one device, showing that the particle had, as expected, passed through only one slit. But as the method of detection is chosen by the experimenter, in a sense the observer decides how he or she wants the particle to behave.

There is a more subtle but enormously significant implication. The difference between the two outcomes reflects the difference in the experimenter's knowledge. When a light-sensitive screen is used to detect a photon, the experimenter has no way of telling which slit it has passed through, so it appears as if it has passed through both, giving a wave-like effect. With telescopes the experimenter *can* tell which slit the photon went through and the photon therefore obligingly acts like the particle it is supposed to be. In other words, it is not just the outcome of the experiment, but the behaviour of the particle itself that seems to depend on what the observer knows – almost as if it depends on the physicist to give it form. When he or she has specific information, the particle behaves specifically; when they have only vague information, the particle behaves vaguely, as if nobody had told it exactly what to do.

In the 1950s Richard Feynman came up with an interpretation of the double-slit experiment based on quantum mechanics that may seem bizarre – even for this strangest of disciplines – but which fits both its theory and practice. According to his interpretation a photon does not take a single path towards the target, but *simultaneously* takes every possible path – it really does go through both slits. The potential paths of the particle represent a series of probabilities, or possibilities, known as a 'wave function'. It is only when the particle is observed that the wave function 'collapses' and the particle takes on a definite position and path. As John Archibald Wheeler, who taught Feynman, explains (his emphasis): 'Each photon is governed by laws of probability and behaves like a cloud *until it is detected* . . . The act of measurement is the transforming act that collapses uncertainty into certainty.'[30] Put another way, until it is measured the photon 'remains an ethereal cloud of probability precisely because it is unobserved'.[31]

If this is correct it would apply to every particle in the universe, and to every property of every particle. They are all wave functions, waiting to receive specific values by being observed. Of course this doesn't mean physicists have a special power that makes subatomic particles submit to their will. What the double-slit experiment and others reveal is the existence of an intimate, and positively spooky, connection between *anyone's* mind and any matter in the universe.

THE MECHANISM OF GENESIS

John Archibald Wheeler (1926–2008) proposed the most far-reaching interpretation of the observer effect. One of the giants of theoretical physics, Wheeler studied under Neils Bohr and Einstein. During the 1930s he worked with Bohr and Enrico Fermi on the theory behind the atomic bomb before then moving on to work on the wartime Manhattan Project. He coined the terms 'black hole' (the existence of which he predicted theoretically) and 'wormhole'. In the 1979 *New York Review of Books*, the mathematician Martin Gardner wrote of Wheeler:

No one knows more about modern physics than Wheeler, and few physicists have proposed more challenging speculative ideas. In recent years he has been increasingly concerned with the curious world of QM [quantum mechanics] and its many paradoxes which suggest that, on the microlevel, reality seems more like magic than like nature on the macrolevel. No one wants to revive a solipsism that says a tree doesn't exist unless a person (or a cow?) is looking at it, but a tree is made of particles such as electrons, and when a physicist looks at an electron something extremely mystifying happens. The act of observation alters the particle's state.[32]

Wheeler made a simple but profound observation about the double-split experiment that took the observer effect to a whole new level. As we have seen, the outcome – particle or wave – depends essentially on how much information the experimenter chooses to have. Wheeler argued this would even apply if the experimenter possessed the information only *after* the experiment had been run.

To demonstrate this he devised the simple 'delayed choice' thought-experiment. Imagine the double-slit experiment was set up so it had *both* a light-sensitive screen and, behind it, two telescopes, one trained on each slit. If the experimenter could somehow decide *after* the photon had passed through the slits which type of detector would come into play then, Wheeler pointed out, logically exactly the same results would apply as if the experimenter had decided what would happen in advance. The screen would show waves, the telescopes particles.

Again, the outcome would reflect the experimenter's knowledge, but this time they would only have this knowledge after the event. So, if in the normal experiment the observer determined how the particle was going to behave, in Wheeler's delayed choice version, they determined how it *did* behave. The observer could decide how a particle behaved in the past, even if only a few microseconds before. As Wheeler pointed out, thinking this process through logically, you come up with backwards causality – time working the 'wrong' way round.

When it was first advanced, the delayed choice experiment could only be an intellectual exercise. After all, how could the experimenter make the decision and throw the switches in the infinitesimally short time the photon was between the slit and the detector, travelling at the speed of light? But in 2006, after many unsuccessful attempts, a means of running this experiment for real was devised. A team of French physicists led by Vincent Jacques used a device that allowed a single photon to take either a single or double path, the choice being determined by a quantum Random Event Generator. In this version the experimenter had to make no choice at all, and just had to gather the information at the end of the test. Needless to say the experiment confirmed Wheeler's predictions absolutely.[33]

The delayed choice experiment showed that observations determine events in the past – but how far back could it go? Wheeler came up with another sequence of arguments that showed that it could also work on a cosmic scale. At the time of writing this has yet to be tested experimentally – but the logic holds up.

A well-known phenomenon in astronomy involves the light from a distant star being bent by a body with a massive gravitational force – say a black hole – between the star and Earth. An effect of this 'gravitational lensing' is that, if the star is immediately behind the black hole, then from Earth we see two images of the star, one either side. Wheeler pointed out that as the light from the star consists of individual photons, this double imaging means that some have been bent round one side of the black hole, and some round the other. Effectively like being passed through the two slits in a lab. If an experimenter on Earth ran the double-split experiment using light from the star, the result should be exactly the same as with the traditional experiment and the delayed choice version: particles or waves depending on how the experimenter chooses to detect them.

Only in this version of the experiment, the light would have been emitted from the star millions, even billions, of years ago. Obviously it would hardly be possible to decide in advance whether they should be particles or waves. So the choice of today's observer would be to decide which side of the black hole the photons would pass, even though it happened many millions of years ago. As Wheeler explains:

Since we make our decision whether to measure the interference of the two paths or to determine which path was followed a billion or so years after the photon started its journey, we must conclude that our very act of measurement not only revealed the nature of the photon's history on its way to us, but in some sense *determined* that history. The past history of the universe has no more validity than is assigned by the measurements we make – now![34]

Paul Davies and John Gribbin comment on the implications of Wheeler's argument: 'In other words, the quantum nature of reality involves non-local effects that could in principle reach right across the Universe and stretch back across time.'[35]

From such reasoning, Wheeler came to a truly extraordinary vision of the role of the mind in the universe. Realizing that the idea that observers affect what they observe only scratches the surface, Wheeler proposed that we should think not in terms of observers but of *participants*. He then asked whether the difference between observation and participation might be 'the most important clue we have to the genesis of the universe':[36]

> The phenomena called into being by these decisions reach backward in time in their consequences . . . back even to the earliest days of the universe . . . . Useful as it is under everyday circumstances to say that the world exists 'out there' independent of us, that view can no longer be upheld. There is a strange sense in which this is a 'participatory universe'.[37]

In a positively *Star Trek*-like sound bite, Wheeler declared that: 'We are participators in bringing into being not only the near and here but the far away and long ago.'[38] From this reasoning he formulated an even more extreme version of the anthropic principle. We saw earlier that this has been conceptually divided between the weak anthropic principle (the universe looks as if it was designed for life but this is probably an illusion) and the strong anthropic principle (the universe *is* designed for life). But Wheeler came up with what he termed the participatory anthropic principle – that *we* are designing the universe. The theory's many knee-jerk detractors were delighted to discover that its acronym is 'PAP'.

According to Wheeler's big idea, the universe was not designed to produce intelligent life for the fun of it; intelligent life is necessary for the universe itself to exist. Writing in 1977 he stated:

> The quantum principle shows that there is a sense in which what the observer will do in the future defines what happens in the past – even in a past so remote that life did not then exist, and shows even more, that 'observership' is a prerequisite for any useful version of 'reality'. One is led by these considerations to explore

the working hypothesis that 'observership is the mechanism of genesis'.[39]

Recognizing the momentous nature of Wheeler's hyptho-thesis, Bernard Carr comments:

> Wheeler has suggested a more radical interpretation in which the universe does not even come into being in a well-defined way until an observer is produced who can perceive it. In this case, the very *existence* of the universe depends on life.[40]

The theory eliminates the need for the multiverse as a solution to the dilemma of the anthropic principle. If the universe needs observers in order to exist then, 'no universe at all could come into being unless it were guaranteed to produce life, consciousness and observership somewhere and for some little length of time in its history-to-be'.[41]

PAP is admittedly an extreme theory. Its potential for being misunderstood and exploited by a whole range of non-scientists including New Agers and science fiction fantasists is only too obvious. Wheeler was particularly incensed that his cosmological ideas were continually used in attempts to explain parapsychological and paranormal phenomena or were even taken to mean that they had already explained them. As a fierce opponent of psi and a board member of the American Association for the Advancement of Science, in 1969 he (unsuccessfully) opposed the admission of the Parapsychological Association as an affiliate member. Ten years later – furious at finding himself speaking alongside parapsychologists at an AAAS conference – he tried to have the decision rescinded, writing a hard-hitting paper entitled 'Drive the Pseudos Out of the Workshop of Science' that he and fellow sceptics circulated widely, again unsuccessfully.

At first it might seem odd that Wheeler took such a line given that his own ideas seem even weirder than the most incontinent claims of the paranormalists. However, his fury was a result of the fact that his interpretation of quantum mechanics has often been twisted to validate unexplained phenomena. Given that Wheeler seems to be saying that the minds of human observers affect the universe at a quantum level, some parapsychologists and many New Agers have taken this to mean that the minds of psychics can, for example, cause changes in the subatomic structure of a spoon, making it bend according to their will alone. Wheeler objected that this was not what he was saying at all.

Wheeler's argument is that by discovering the laws of physics that make the universe tick, sentient observers were and are bringing them into being. But they are not actually *making* the laws. There is no free choice involved. In the double slit experiment, for example, the experimenter can 'make' the particles behave as a particle or a wave, but not as anything else. And whatever the observer is doing to influence behaviour is entirely unconscious. Such experiments show that mind and matter are intimately connected, but in a circular relationship where neither has the upper hand. It's not a case of mind *over* matter, or even matter over mind – both are acting as part of the same process.

One of the intuitive difficulties with the idea of a designer universe is the notion that building a whole universe just to populate odd corners of it with intelligent beings seems rather excessive. Surely GUD could have found a more economical way to work? But Wheeler argues it makes perfect sense if we think not in terms of size but *time*. The universe has to be as big as it is, and to have existed for as long as it has, for the conditions required for life to have arisen. The size and age of the universe are directly related:

if the universe contained only enough matter to make one galaxy, it would not be able to exist long enough to make life. (In fact, Wheeler calculated that a galaxy-sized universe would only exist for about a year.) [42] Barrow and Tipler observe that certain of Teilhard de Chardin's arguments supporting his contention that the purpose of the universe is to produce life are 'strikingly similar to Wheeler's idea that the Universe must be at least as large as it is in order for any intelligent life at all to exist in it.'[43]

Even more relevant to this present discussion, Wheeler relates his theory of the participatory universe to the work of Leibniz, one of his scientific and philosophical heroes. In so doing, Wheeler is therefore, however unknowingly, linking his theory to the Hermetic vision. In an article written in 1970, 'Beyond the Edge of Time', he suggests that the weak anthropic principle 'may only be a halfway point on the road toward thinking of the universe as Leibniz did, as a world of relationships, not a world of machinery' and asks:

> Does the universe . . . derive its meaning from 'partici-pation'? Are we destined to return to the great concept of Leibniz, of 'pre-established harmony' . . . before we can make the next great advance?[44]

John Wheeler is by no means the only eminent physicist to accept such an apparently outrageous idea that we – and all the other intelligent species in the universe – are actually creating the universe, not only now but also back at its beginning. Stephen Hawking, along with collaborators such as American physicist James Hartle and Thomas Hertog of CERN, have embraced much the same idea, and for many of the same reasons. They, too, take the implications of the double-slit experiment and other paradoxes of the quantum world and apply them on a cosmic scale. The major difference with Hawking's vision is that he embraces the multiverse, and so accepts that there are many other universes in which conditions do not support life. Quite how these universes are supposed to exist without benefit of observers is something that is left open.

In his work with Hartle, Hawking extended the idea of wave functions to the entire universe, devising a mathematical formulation – the 'Hawking-Hartle state', developed from one of Wheeler's equations – to express it. Just as the experimenter in the laboratory collapses the wave function of a photon in the double-split experiment, so observations of the universe collapse its wave functions – not only now but in the past. Backward causality, in other words.

In *The Grand Design* Hawking argues that the traditional 'bottom-up' approach to the history of the universe is wrong. Instead of starting with the big bang and working forwards, extrapolating the laws of physics to work out why the universe now is the way it is, we should take a 'top-down' line, working backward from the present. This would allow for the fact that, building on the work of Feynman and Wheeler, our existence now determines how the universe began and evolved: 'We create history by our observation, rather than history creating us.'[45] Or as *New Scientist* put it in a report on Hawking and Hertog's recent work: 'A measurement made in the present is deciding what happened 13.7 billion years ago; by looking out at the universe, we assign ourselves a particular, concrete history.'[46]

Although the comparison would no doubt have truly appalled Wheeler and probably wouldn't find favour with Hawking, such ideas chime very well with the global coherence effect found by Dick Bierman and the Global Consciousness Project. This shares with the participatory universe hypothesis the basic idea that mind is intimately bound up with matter – indeed, that the mind is even a property of matter. Both show that the very presence of thinking entities affects the physical universe at a quantum level.

ARE WE GOD?

This idea of the participatory universe understandably fuels all manner of speculation. Perhaps, as humans observe more and more of the universe, both on a cosmic scale and at a quantum level, the relationship between our consciousness and the universe is becoming more and more interdependent. Perhaps, too, as Teilhard de Chardin thought, we, along with extraterrestrial races, are evolving into a cosmic consciousness. This was the plan all along: in the end, we will all *be* the universe. If this is the case then humans are or will be God, the creator of the universe in the first place.

Or maybe there is a hierarchy of observers, with more advanced beings already taking a more active role in shaping the cosmos. Barrow and Tipler describe a possible extrapolation of Wheeler's vision:

> That there is some Ultimate Observer who is in the end responsible for coordinating the separate observations of the lesser observers and is therefore responsible for bringing the entire Universe into existence.[47]

If so, they speculate, the unfolding of the universe is leading up to the Ultimate Observer's Final Observation, when the cosmic plan will be complete.

Others still seek to keep a more-or-less traditional God in the picture. Keith Ward, British theologian, philosopher and born-again Anglican minister, takes Wheeler's participatory anthropic principle to a new, if predictable, extreme. Ward suggests that it is not humans and extraterrestrials doing the observing and creating: 'God is the ultimate observer or consciousness which creates the reality.'[48] However, he does accept that human consciousness makes a small contribution to the shaping of the universe. But even that small contribution represents a huge leap for a born-again Christian priest.

Despite being speculative, all three extrapolations agree that intelligent, conscious beings – such as humans – are in some way partly the creator.

As we have seen throughout this book, understanding God was one of the central inspirations for science. Isaac Newton, for example:

> ... strove for a unified solution that would encompass not only the mysteries of celestial and terrestrial physics, but also the perennial religious problem of the relation between the Creator and his creation.[49]

Echoing this, the man who is in many ways Newton's modern equivalent, Stephen Hawking, writes in the memorable phrase that concludes *A Brief History of Time* (1988) that the ultimate goal of science is to 'know the mind of God'.[50]

In fact, the quest for the mind of God may effectively be over. In the end, the journey was not a long one and the destination has proved much closer to home than anyone could have imagined. We all have a share in God's mind simply by being human.

The Hermetic quest also sought primarily to understand the mind of God through knowledge of the cosmos, as can be seen from Treatise XI of the *Corpus Hermeticum*, in which Mind explains to Hermes:

> So you must think of god in this way, as having everything – the cosmos, himself, <the> universe – like

thoughts within himself. Thus, unless you make yourself equal to god, you cannot understand god; like is understood by like . . . Having conceived that nothing is impossible to you, consider yourself immortal and able to understand everything, all art, all learning, the temper of every living thing . . . And when you have understood all these at once – times, places, things, qualities, quantities – then you can understand god.[51]

The Hermetic cosmos itself is also described as a thought of God's, the product of his mind – in a sense, his mind itself.

Even if John Wheeler and other celebrated scientists such as Stephen Hawking are not aware of it, the universe they describe has such close parallels with the Hermetic vision – the solar child of the ancient Egyptian religion of Heliopolis – that they might as well be the two encircling strands around the same caduceus of wisdom.

In Wheeler's participatory universe, the consciousness of observers is embedded in its structure and both mind and universe are therefore shaped by and shape each other. We are, or at least are part of, the creative force. If for creative force we read God – and the distinction is only a matter of semantics or personal taste – then essentially all humanity is divine or at least an integral part of the divine.

The creative force and the material universe are locked in an eternal embrace or endless creative waltz. Shifting the terminology again, God is the universe, and vice versa. Intelligent beings are part of God, and also, as their minds help shape the universe, they enjoy a special role in creation. Creator, created and creation are constantly circling in a dazzling dance of ultimate meaning and purpose, an endless jump of joy.

Yet as encapsulated in the Hermetica, this apparent welter of transcendentalism has not been the Holy Grail of many of the world's most brilliant minds simply because they liked the mysticism and poetry – although that certainly has its own appeal. To the Hermeticist, pursuing any intellectual endeavour without including the idea of God would be simply absurd. Very succinctly Treatise XI describes the all-pervasive divine: 'God makes eternity; eternity makes the cosmos; the cosmos makes time; time makes becoming.'[52]

Glenn Alexander Magee writes of the 'Hermetic doctrine of the "circular" relationship between God and creation and the necessity of man for the completion of God'.[53] According to the Hermetica then, humankind has a special place in God's creation. God needs human beings to exist because we are part of God. And we also need God, we need worship, we need awe. The concept of ordinary, everyday humanity in some very real way actually completing God is anathema to, for example, Catholicism, with its fixation on sin, purgatory and subservience to priests and a deity whose separate being is always above and beyond us.

Wheeler says essentially the same thing as the Hermetica: the circular relationship between mind and the universe makes human consciousness necessary for the completion of the universe. The same idea is found in Neoplatonism, which is hardly surprising given the Egyptian roots it shares with Hermeticism, through its founder Plotinus, student of the mysterious Egyptian sage Ammonius Saccas. As Magee notes: 'Like the Hermeticists, Plotinus holds that the cosmos is a circular process of emanation and return to the One'. [54]

From the same basic reasoning as our own – which is based on the growing evidence of design and purpose being uncovered by all the sciences – Austrian astrophysicist Erich Jantsch argued in the 1970s that the universe was 'self-organizing': 'God is not the creator, but the mind of the universe.'[55] Although Jantsch found this concept behind many of the great mystical religions, one lay behind them all. He explains that, 'In the oldest recorded world view, Hermetic philosophy . . . this wholeness resting in itself is called the "all".'[56] Jantsch seems to here recognize the origins of Hermeticism in the religion of Heliopolis, whose Pyramid Texts are indeed the world's oldest cosmological writings.

The same matrix of connections exists between the picture emerging from quantum physics and Heliopolitan thought. In Wheeler's system, the laws of physics build the material universe, which eventually gives rise to living organisms, which eventually produce sentient beings able to observe and understand the cosmos. By discovering how the universe works, observers are actually creating it in the far distant past – even at the beginning of time. Wheeler saw this as a cycle or feedback loop whereby the universe creates sentient beings who then return the loop back to the beginning. He encapsulated this cycle in his famous diagram showing the eye of the observer looking back at the beginning of the universe (see illustrations) and in the words:

> Beginning with the big bang, the universe expands and cools. After eons of dynamic development it gives rise to observership. Acts of observer-participancy . . . in turn give tangible 'reality' to the universe not only now but back to the beginning.[57]

Exciting though this may be, what Wheeler describes is by no means a new concept. It resoundingly echoes key ideas of the Pyramid Texts, which speak of how Atum created the big bang – very literally – giving rise to the expanding and ever-more complex universe that ultimately created people, who live on the edge of manifestation, in what Karl Luckert calls the 'turn-around realm', the inner place where human consciousness begins its return journey to Atum. And it is not just to him that human consciousness returns, but to his very act of creation – in other words, back to the big bang.

In a deeply satisfying exchange, not only does the latest scientific thinking support the Hermetic cosmology but Hermeticism in turn makes sense of the discoveries of science . . . This is as it should be, for it was a brutal operation that severed the two. And now they seem to be calling to each other like separated twins, aching to be as one again.

CHAPTER THIRTEEN

ESCAPING FROM FLATLAND

The Hermetica should, at the very least, be given its due because of its truly towering influence over our culture and history since the fifteenth century, especially its powerful role in creating science – though today's practitioners themselves are either unaware of or unwilling to accept this fact. As Richard Westfall writes in relation to Newton:

> The Hermetic elements in Newton's thought are not in the end antithetical to the scientific enterprise. Quite the contrary, by wedding the two traditions, the Hermetic and the mechanical, to each other, he established the family line that claims as its direct descendant the very science that sneers today uncomprehendingly at the occult ideas associated with Hermetic philosophy.[1]

This convergence of the mechanistic with the mystical is recognized, albeit apparently unconsciously, by the likes of Wheeler, who repeatedly related his work to Leibniz – in turn, at the very least a closet Hermeticist whose own hero was Giordano Bruno – writing, for example:

> Inspect the interior of a particle of one type, and magnify it up enormously, and in that interior see one

view of the whole universe (compare the concept of monad of Leibniz (1714), 'The monads have no window through which anything can enter or depart'); and do likewise for another particle of the same type. Are particles of the same pattern identical in any one cycle of the universe because they give identically patterned views of the same universe? No acceptable explanation for the miraculous identity of particles of the same type has ever been put forward. That identity must be regarded, not as a triviality, but as a central mystery of physics.[2]

Westfall points out that the term 'occult' first took on its negative connotation when seventeenth-century mechanistic scientists began to use it as a putdown. And so the golden age of scientific mystics was brought down to the level of the sinister, illusory, cheap and nasty. But in fact, 'occult' was originally a synonym for 'Hermetic'.[3]

After immersing ourselves over the years not only in the history of religions and heresies but also in the history of science, in talking to scientists, delving into the obvious and less obvious learned papers and attending lectures from the very abstruse and arcane to the most direct mechanistic science, we have concluded – along with many of those we quote in this book – that science still needs the Hermetic wisdom.

Science would have found it considerably easier to make sense of the data that it is now uncovering – the designer universe, life as a cosmic imperative, the directionality of evolution, the participatory universe – if it had never jettisoned the Hermetic framework. In fact, it would have *predicted* these discoveries. And although it is impossible to know for sure, we believe the signs are there in the texts themselves that a Hermeticized science would have already advanced far beyond the point that we have reached today.

But all is not lost. David Fideler, editor of *Alexandria: the Journal of Western Cosmological Tradition,* argues that modern science is moving ever more in a Neoplatonic (for which read Hermetic) direction:

> Over the last century the mechanistic view of the universe has started to completely break down. Because the implications of quantum mechanics, chaos theory, and the realization that we inhabit an evolutionary, self-organizing universe are starting to work themselves out, it is no exaggeration to say that we are truly living in the midst of a new Cosmological Revolution that will ultimately overthrow the Scientific Revolution of the Renaissance. And if the mechanistic world view left us stranded in Flatland – a two dimensional world of dead, atomistic matter in motion – the emerging cosmological picture is far more complex, multidimensional, and resonant with the traditional Neoplatonic metaphor of the living universe.[4]

Is the 'living universe' merely a metaphor? Was it ever? Hermeticists certainly meant it literally. Yet humanity *is* stranded in 'Flatland', shut off from the radiance of the Hermetic vision and all the vast benefits it bestows. This, however, is not inevitably humanity's end. We can – and must – escape from Flatland.

Fideler refers to the holistic nature of existence, citing the fact that in 1982 physicists showed particles of light from a common source 'continue to act in concert with one another' no matter how far apart they are, a phenomenon known as 'quantum nonlocality'. He explains:

> The tantalizing implication of quantum nonlocality is that the entire universe, which is thought to have blazed forth from the first light of the big bang, is at its

deepest level a seamless holistic system in which every 'particle' is in 'communication' with every other 'particle', even though separated by millions of light years. In this sense, experimental science seems to be on the verge of validating the perception of all mystics – Plotinus included – that there is an underlying unity to the cosmos which transcends the boundaries of space and time.[5]

Fideler argues that the breakdown of the mechanistic worldview requires a new type of science, and proposes that a fusion of the philosophy of Plotinus and Wheeler's concept of the participatory universe should provide the model. The consequence, says Fideler, is that:

> . . . the focus of life will become more multidimensional, contemplative, and celebratory as we as individuals come to see ourselves as living embodiments of the-universe-in-search-of-its-own-Being, and as active participants in the ongoing creation of the world.[6]

Unsurprisingly, the ancient source of both Neoplatonism and Hermeticism, the wisdom of Heliopolis, also offers a way forward, out of Flatland. Karl Luckert states emphatically:

> Logic is not abandoned when one tries to understand human existence the ancient Egyptian way; namely, from the perspective of divinely radiated energy and life, from within emanations of divine purpose and pleasure, or from sun rays which in turn engender what we, nowadays, regard as being more 'substantial' protoplasm and genes . . . Eternity itself will arbitrate between moribund analytic and disjunctive reasoning, on one hand, and the type of holistic reasoning which was cherished by Heliopolitan priests on the other.[7]

Yes, science should undoubtedly be more contemplative, inviting practitioners to utilize every level of their minds without embarrassment or shame. The subconscious mind, usually quiescent under a welter of reason and mundane concerns, has long been acknowledged as the most fertile repository of inspiration and even otherwise hidden knowledge. Take the famous case of the German chemist August Kekulé (1829–1896), who, together with a great many of his scientific peers, had been puzzling for a long time about the structure of benzene, but without success. Falling into a daydream or reverie he saw a snake swallowing its own tail. On coming back to normal consciousness, he realized he had been presented with the answer: six carbon atoms in a ring . . . This was not his only example of subconscious prodding. On an earlier occasion a reverie had also provided him with crucial information. On the top of a London omnibus an image of dancing molecules floated into his head, giving him the insight into the theory of chemical structure – and securing him a place in scientific history.

Backed up by arduous study and hard facts, the use of intuition and hunches often provide similar short-cuts – if they are allowed to. Had Kekulé dismissed his insights as 'just daydreams' he might never have made his great discoveries.

As that episode reveals, the subconscious mind deals in symbolism and poetry – hence the distinctive surrealism of dreams – the very language that enables the Hermetic texts to seduce and penetrate all levels of the mind at once. Such symbolism is not moonshine or mumbo jumbo. It is a direct message to the centre of every mind.

THE NEW SCIENCE
The history of science portrays the mechanistic revolution as an inevitable coming to our senses, a right and proper intellectual maturation. But the reality is that the move

away from the mystical side of science was a historical accident. James I's paranoid hatred and fear of witches made it expedient for the likes of Francis Bacon to be seen to have no occult connections, so that side of Hermeticism rapidly became not only unwise, but unfashionable. And the Counter Reformation made it equally dangerous for non-Catholics to be occultists (Catholic occultists not being terribly welcome either), while the French Catholics built up Descartes to oppose the despised Rosicrucianism. If events in the seventeenth century had been slightly different, no doubt all our science would have continued to work undisturbed within the Hermetic principles right through to today. After all, with such a distinguished track record it would have been foolish to junk it for no reason.

And if the Hermetica had remained influential in academia, science is not the only field that would be different, since the understanding of the universe it bestows affects pretty much everything else in our culture.

When accepting the Liberty Medal on 4 July 1994, Václav Havel, the former dissident playwright who became the first President of the new Czech Republic after the end of the Cold War, lamented the way human rights and freedoms, despite all the big changes that came with the downfall of communism and end of the Cold War, had become 'mere froth floating on the subsiding waters of faith in a purely scientific relationship to the world'.[8] He went on to say that:

> Paradoxically, inspiration for the renewal of this lost integrity can once again be found in science. In a science that is new – let us say post-modern – a science producing ideas that in a certain sense allow it to transcend its own limits.[9]

Havel cited as examples of this 'post-modern science' both

the anthropic cosmological principle and the Gaia hypothesis. Of the anthropic principle he said:

> This is not yet proof that the aim of the universe has always been that in a certain sense it should one day see itself through our eyes. But how else can this matter be explained?[10]

In his view the anthropic principle shows that 'we are mysteriously connected to the entire universe; we are mirrored in it just as the entire evolution of the universe is mirrored in us'.[11]

If science had been uninterruptedly Hermetic, would the environment be in the same terrifying condition we find it in today? Almost certainly not. Without over-sentimentalizing, the Earth itself would have been cherished as a living being. There would be no question of having to fight for human rights or the right of animals to be treated gently and with respect. If every human and every beast is an integral part of all creation, then they are all part of us in a very real way. Hurting them would be hurting ourselves. The Hermetic system adds a moral centre to science, which is largely lacking in its amoral mechanistic manifestations and depends almost entirely on the ethics and integrity of individual practitioners.

We began this book by arguing that the magical world-view is essentially hardwired into humanity. Now we can see this is because human beings are aware, at some deep level, of the true nature of the universe and our astonishingly significant role in it. We are indeed hardwired to feel the hollowness of the God-shaped hole deep inside, as the Hermetica acknowledges: 'Praising god is in our nature as humans because we happen to be in some sense his descendants . . .'[12]

The evidence that science itself has produced supports

the essential ideas that underpin the sense of Otherness innate to human beings. Inconvenient though it may be for the Dawkins' school, there is no doubt that cosmology, physics and many other disciplines, including even biology, present evidence that the universe is non-random, meaningful and designed for life. Science has even felt compelled to rewrite its own rules when it comes across evidence of purpose and design, as is evidenced by the overzealous embracing of the multiverse. It is as if the scientific world is terrified that admitting *anything* non-random will let all the religious 'nonsense' back in.

As with any philosophy worth contemplating, it is the implications that really matter. The path of Hermes Trismegistus illuminated the radiant Renaissance spirit, which burst forth from Pico della Mirandola's *Oration on the Dignity of Man*, which with its high praise for 'miraculous man' cleared with one bound the bigot-built walls that imprisoned human ignorance. Human beings are brilliant because we are all potentially gods and creators. Not born in sin and dirt but in joy and brightness, entering the world not as devil-filled infants but in William Wordsworth's famous words 'trailing clouds of glory'. The implications of being god-like humans are enormous. Nothing is beyond us. We can literally reach for the stars. As the Hermetica emphasizes:

> For the human is a godlike living thing, not comparable to the other living things of the earth but to those in heaven above, who are called gods. Or better – if one dare tell the truth – the one who is really human is above these gods as well, or at least they are wholly equal in power to one another.[13]

Likewise, Plotinus wrote of 'finding the strength to see divinity within'.[14]

However, the Hermetic impetus to find new worlds to conquer carries with it a sense of responsibility. True Hermeticists can never be dictators nor seek to crush the weak and the vulnerable. For if they themselves, as they believe, are also the universe and even God, how can they damage a fellow god in need of their help? As the *Corpus Hermeticum* states profoundly: 'There is but one religion of god, and that is not to be evil.'[15]

In the 1970s there was a vogue for books linking the discoveries of physics with Eastern mysticism, such as the works of Fritjof Capra, which provided many seekers with some degree of nourishment to assuage their spiritual hunger. But we should acknowledge that the West has its own, forgotten tradition – Hermeticism – just waiting to provide comfort, knowledge, excitement and freedom.

Like any idea that can turn the world around, the Hermetic universe has been forbidden by the powers of intellectual darkness. The Church demonized it, fearing its potential for firing up generations of men and women to think for themselves about any subjects that seized their hearts and minds. And after science disowned and disinherited it, originally out of expediency, it became an ingrained prejudice. But the Hermetic flame never died and now, thanks to science itself, the fire – in all but name – seems ready to erupt into the world.

If any one individual symbolizes the tormented history of the Hermetic tradition it is Giordano Bruno. Although a rather sinister statue now stands in Rome at the site of his execution, providing a focus for crowds of pilgrims, few of them seem to realize exactly what he died for. Poor Bruno is either completely ignored or totally misunderstood – if he is remembered at all. He is ultimately portrayed as condemned by the Church either for preaching the existence of the infinite universe or for his support for Copernicus. In a 2010 Reith Lecture, Lord Rees said: 'The

Italian monk and scholar Giordano Bruno, burnt at the stake in 1600, conjectured that the stars were other "suns", each with their retinue of planets.'[16] The implication is that he died for science in the modern sense. But Bruno was, in reality, a martyr for *the Hermetic tradition*.

In Europe, the Church told their flock that they were individually weak, miserable sinners, but then the Hermetic Renaissance declared they were quite the opposite, lighting the way to the scientific revolution. In the beginning all science was Hermetic science. But something went badly wrong. When it junked the Hermetic philosophy, science began to preach that we owe our existence to a long series of accidents and that ultimately our lives have no meaning. The sense of unlimited horizons and the joy of being alive were eroded.

When the scientific wisdom was plucked from Hermeticism to fuel the engines of progress for today's world and the underlying transcendentalism rejected, the whole tradition lost its soul – specifically the feminine aspect of its soul. When science set its stern face towards the test tube and the slide rule it was in effect turning its back on *Sophia*, the female aspect of the Hermetic knowledge, literally God's other half. And in the ironic replay of the excision of the sacred feminine from Christianity, here science lost not only its soul but also its heart.

Although the names of the great Hermeticists that have come down to us are resolutely male, practitioners such as Bruno took pains to emphasise the rightful place of the feminine, of Isis and Sophia, in the great scheme of things. We suggest that this was not merely some poetic turn of phrase, but a profound acknowledgement of the necessity to embrace the female side of learning and understanding. Whereas men tend to be literal and logical, women tend to think in much more holistic and symbolic ways. To most women who understand the divine, it can be understood immediately, as a whole. It is not necessary to spell things out or limit their participation in the cosmic dance with hard dogma and punishment. That is what terrified the Inquisitors, and what continues to disturb the Church authorities today.

To be a Hermeticist, no matter what one's gender, is to accept and utilise both male and female mindsets, embodied in the ancient Hermetic and alchemical symbol of the hermaphrodite. Only by becoming whole oneself can the universe be finally understood and totally participated in.

But science, like the Judeo-Christian religions, severed its ties with Sophia, with its other half. And although it can weigh, measure, calculate and send men to play golf on the moon, the real awe and glory of the universe lies in the human heart and soul. If it is allowed to be whole. This was Bruno's message. This was the ancient wisdom. And simple though it may seem, it is in itself one of the profoundest secrets of all.

The moment to restore the sense of wonder is long overdue. There has never been a better time to let the 'miracle of man' back in.

APPENDIX

HERMES AND THE FIRST HERETIC

Sometimes research turns up exciting connections that frustratingly don't belong to the main argument of a book. As some of the information we uncovered on the origins of the Hermetica isn't directly relevant to *The Forbidden Universe* but relates to unfinished business in our previous book, *The Masks of Christ*, we have included it in this appendix.

The inclusion of Hermetic texts such as a Coptic copy of *Asclepius* in the famous collection of books discovered at Nag Hammadi in Egypt in 1945 (often referred to as the Gnostic Gospels) revealed the close connection between Gnosticism and Hermeticism. Brian P. Copenhaver explains its significance (his emphases):

> The impact of the Nag Hammadi discoveries on our understanding of the *Hermetica* has been enormous. To find theoretical Hermetic writings in *Egypt*, in *Coptic* and alongside the wildest efflorescences of the *Gnostic* imagination was a stunning challenge to the older view . . . that the *Hermetica* could be entirely understood in a post-Platonic Greek context.[1]

Other Nag Hammadi books may be largely innocent of the 'wildest efflorescences' but they do have 'doctrinal parallels'[2] with the Hermetica. Although this shows that the writers came from a similar school, they often extrapolated their ideas very differently, sometimes in strangely incompatible ways. (Plotinus wrote a tract called *Against the Gnostics*, accusing them of developing their ideas erroneously.)

The discovery had a major impact, and went so far as inspiring the classic *The Gnostic Religion* (1958), by the German-American philosopher Hans Jonas, to discuss Hermeticism alongside the more familiar Gnostic systems.[3]

Thanks to Dan Brown's blockbusters, millions of people worldwide now know about Gnosticism, the version of Christianity that was eventually anathematized by the emergent Catholic Church and which is associated most with what the Church would have concealed from us. (One of the main revelations of the Nag Hammadi books was the importance of Mary Magdalene and her apparently intimate relationship with Jesus.)

The precise origins of Gnosticism are uncertain and controversial. In a religious sense the term first surfaces towards the end of the second century CE in a Christian context, referring to a sect deemed heretical by the early Church because of its different view of God, Jesus and the path to salvation. The word itself derives from the Greek *gnostikos*, which simply means the ability to acquire knowledge. These heretics called themselves *gnostikoi* – 'knowers' – but the term was also applied to many similar Christian sects, each with its very different views.

The essential difference – what really set them beyond the pale to their detractors – was that these sects believed an understanding of God and individual salvation could be won through direct *personal* experience. Furthermore there was no need for a Church or priesthood as intermediaries – which posed an obvious challenge to the power of the Vatican, with its emphasis on faith rather than under-standing, and on collective experience.

Until the last century or so, the earliest known accounts of Gnosticism were found in hostile Christian writings, which stated it grew out of Christianity and therefore post-dated Jesus and Paul. However, more recent research has revealed that Gnostic beliefs were not confined to Christianity, and that the Christian Gnostics had drawn their worldview from earlier pagan sources, adapting them to the teachings of Jesus.

As a result, the question of the origins of Gnosticism has been hotly debated ever since, but without reaching any conclusive answer. What is known is that it first appeared in the Middle East, particularly Egypt. Different historians champion a Greek, Jewish or Iranian background, or a fusion of all three in Hellenic Alexandria. But once again it is Egypt that beckons.

The fundamental problem in attempting to trace Gnosticism to its source is that there is no agreed definition of 'Gnostic'. To non-specialists (and New Agers) it simply refers to the attitude that salvation or enlightenment is in one's own hands, and requires personal communion with the divine. For academics it describes a specific set of beliefs about the nature of the material world. But there is no consensus about what they are. Even the accepted definition varies between different countries.[4] That being said, they do agree on certain basic facts.

Gnostics see the material world as inherently flawed, separated from its creator, and believe that the divine and material are mutually antithetical, a belief known as dualism. For Gnostics, salvation is escaping from the prison of the material world, although different Gnostic sects came up with wildly different ways of doing so. For the Christian Gnostics, this meant devising a radically different interpretation of the nature and role of Jesus from the one held by the early Church – another reason why it hated them. (Whether the Church was wrong and the Gnostics right is sadly outside the scope of this book.) Another defining characteristic of Gnosticism is a belief that the god of this world isn't the *real* God. A Kafkaesque, and even *Matrix*-like sense of illusion permeates much of Gnostic thinking. This is hardly a coincidence: *The Matrix* movies unashamedly draw on Gnostic ideas.

Different Gnostic schools veered off in different directions: the god of this world may be acting under the true God's instructions, may be an evil entity masquerading as God or may be deluded into believing that they actually are God. Then there is the question of the practical applications of spirit-matter dualism: it might lead to asceticism and mortification of the flesh, as it often did. Or it might lead into hedonistic indulgence in the world of the senses – as indeed it did also.

But the parallels with Hermetic and Neoplatonic (and for that matter Platonic) thinking are striking. Neoplatonist belief in the Demiurge and theurgy are essentially the same as that of Gnosticsm, as are the Hermeticists' belief in the 'second god' and the possibility of enlightenment through direct communion with the divine.

Excitingly for us, especially considering our conclusions in *The Masks of Christ*, the clearest signs of Egyptian influence are unequivocally right there in the writings of the man the Church declared the 'first heretic', the Samaritan Simon Magus, a contemporary of Jesus.[5] The extraordinarily colourful Magus is regarded by modern scholars as a 'proto-Gnostic' rather than a Gnostic proper, blending ideas from which Gnosticism, according to the standard definition, was to emerge.

This is Simon's own summary of his theology:

There is one Power, divided into upper and lower, begetting itself, increasing itself, seeking itself, finding itself, being its own mother, its own father . . . its own daughter, its own son . . . One, the root of All.[6]

As Karl Luckert points out, this belief system is strikingly similar to that of the distinguished priests of ancient Heliopolis, revealing yet again their presence throughout history and their resurgence in the early centuries CE.[7] But in the context of Simon Magus we also see how it underpinned even – perhaps especially – the Samaritan religion.

The revelation of an intense kinship between proto-Gnosticism and the Heliopolitan/Hermetic tradition was frankly music to our ears. In *The Masks of Christ* we argue that the exercise of comparing Simon Magus with Jesus strangely elucidates many of the key mysteries and paradoxes about the life and mission of Christ. Although this is the last thing Christians want to hear, the two men were so similar – embodying the same paradoxical blend of the Judaic and pagan – that Simon threatened to undermine Jesus' special status. As a result, the early Church literally demonized him. But according to very early Christian sources, the two men even shared John the Baptist as teacher. Astonishingly, the evidence is that John chose, of all people, Simon Magus as his successor – and that the headquarters of the Baptist's sect were in Alexandria.[8]

We concluded that the explanation of Jesus' mission lies with the Samaritans, who preserved a more faithful version of the original Israelite religion, and which both Simon Magus and Jesus – as well as John the Baptist – were attempting to restore to all the peoples of Israel, including the then-dominant Judeans, or Jews as they became known. But Luckert's identification of a common thread between Simon Magus' theology and ancient Egypt raises certain basic questions with remarkably far-reaching implications. What does the Samaritan link mean for the history of

Christianity? And what does it imply about the true significance of Hermeticism?

If the teachings of Simon Magus were ultimately derived from Heliopolis, this would not only suggest that John the Baptist shared that legacy, but a very real connection with the Hermetica also emerges. So perhaps it is significant that the Dutch theologian and historian Gilles Quispel, one of the editors of the Nag Hammadi texts, writes:

> Owing to the new Hermetic writings that were discovered near Nag Hammadi in 1945, it has become certain that the Hermetic Gnosis was rooted in a secret society in Alexandria, a sort of Masonic lodge, with certain rites, like a kiss of peace, a baptism of rebirth in spirit and a sacred meal of the brethren.[9]

At the very least this connection reinforces the beliefs of the Renaissance Hermeticists, as expressed most robustly by Giordano Bruno, who also considered Jesus to have attempted to return Judaism to its Egyptian roots. Bruno taught that Jesus practised Egyptian magic. Partly based on the comparison with Simon Magus and partly on other historical evidence, in *The Masks of Christ* we argue that Jesus was perceived in his own time primarily as an Egyptian-style magus.[10]

These links are both exciting and tantalizing, and offer golden opportunities for yet more profound, even in their own way sensational, discoveries to be made about Egypt's true legacy to the intellectual, emotional and spiritual life of the West.

NOTES AND REFERENCES

Introduction
1 Quoted in Leake and Sniderman.
2 Quoted in *ibid.*
3 Dawkins, *The God Delusion*, pp. 200–8.

Chapter One
1 Morris A. Finocchiaro, from his introduction to Galileo, *Galileo on the World Systems*, p. 2.
2 Davies, *The Goldilocks Enigma*, p. 147.
3 Our translation of the Latin: *'Siquidem non inepte quidam lucernam mundi, aln mentem, aln rectorem vocant. Trismegistus visibilem Deum . . .'*
4 For example, Washington State University's World Civilizations website: www.wsu.edu:8001/~dee/REN/PICO.HTM
5 Pico della Mirandola.
6 *Ibid.*
7 *Ibid.*
8 *Ibid.*
9 *Ibid.*
10 See Yates, *Giordano Bruno and the Hermetic Tradition*, pp. 87–91.
11 Some academics prefer 'Hermetism', while others use that term for the original philosophy of the early centuries CE and 'Hermeticism' for its Renaissance reincarnation.
12 Tuveson, p. 9.
13 E.g. the opening of Treatise XVI (Copenhaver, p. 58).
14 Lindsay, p. 166.
15 Tuveson, p. xi.
16 Magee, p. 10.
17 Copenhaver, p. 36.
18 Magee, p. 9.

19 Copenhaver, p. 69.
20 Tuveson, p. xii.
21 The relationship between the Sabians of Harran and the Sabians mentioned in the Qur'an – known to us today as the Mandaeans, a baptismal sect whose homeland is in southern Iraq and Iran and who venerate John the Baptist as their great teacher – is a matter of controversy. The line taken by the Arab chroniclers who first set down the al-Mamun story – the earliest account was written about a hundred years after it was supposed to have happened – is that the Harranians took the name simply because although it appears in the Qur'an by then everyone had forgotten who the Sabians were. This is also the position of most historians. However, there is an intriguing complication, as the Mandaeans also have an ancient link with Harran, which seems to be stretching coincidence rather far, especially for us personally since they were central to our research on the true status of John the Baptist, as discussed in our books, *The Templar Revelation* (Chapter 15) and *The Masks of Christ* (Chapter 7).
22 Gündüz, pp. 157–8 and 209.
23 *Ibid.*, p. 208.
24 Churton, *The Golden Builders*, p. 27.
25 See *ibid.*, p. 38.
26 E.g. Copenhaver, p. xlvi.
27 Tuveson, p. ix.
28 Parks, p. 207.
29 Tompkins, p. 52.
30 Yates, *Giordano Bruno and the Hermetic Tradition*, p. 7, quoting an 1871 translation by William Fletcher. Copenhaver (p. 71) renders the phrase as 'progeny of his own divinity'.
31 Copenhaver, p. 2.
32 *Ibid.*, p. 89.
33 E.g. in *Asclepius* (*ibid.*, p. 85).
34 *Ibid.*, p. 59.
35 *Ibid.*, p. 61.
36 Yates, *Giordano Bruno and the Hermetic Tradition*, pp. 154–5.
37 Churton, *The Golden Builders*, p. 59.
38 1 Chronicles 16:30 (TNIV).
39 Joshua 10:12–13 (TNIV).
40 Kepler, p. 391.
41 *Hamlet*, Act II, scene 2.
42 *Ibid.*
43 *Ibid.*
44 Gingerich, p. 23.

45 See Couper and Henbest, pp. 111–3.
46 Quoted in *ibid.*, p. 116.

Chapter Two
 1 Arianism was an alternative view of the nature of Christ that had been rejected and condemned during the formative years of the Catholic Church in the fourth century. In contrast to what became the Church's official position – that God and Christ were of the same substance and that Christ had co-existed with God from the beginning of time – Arianism held that God had created Christ at a specific moment in time. This made him something more like the Gnostic Demiurge – or Hermes' 'second god' – implying that Christ was distinct from God and that there was a time when he had not existed. The Arian view, contrary to a common misconception, was not that Jesus was a mortal chosen by God.
 2 Yates, *Giordano Bruno and the Hermetic Tradition*, p. 11.
 3 Copenhaver, p. 83.
 4 See Picknett and Prince, *The Masks of Christ*, pp. 371–81.
 5 Quoted in Yates, *Giordano Bruno and the Hermetic Tradition*, p. 340.
 6 *Ibid.*, p. 215.
 7 Quoted in *ibid.*, *Giordano Bruno and the Hermetic Tradition*, p. 204.
 8 Quoted in *ibid.*, p. 206.
 9 *Ibid.*, p. 288.
10 See our *The Masks of Christ*, pp. 197–201 and 222–4.
11 Yates, *Giordano Bruno and the Hermetic Tradition*, p. 211.
12 Quoted in *ibid.*, pp. 281–2.
13 Quoted in Tompkins, p. 75.
14 Atanasijevic, p. xxiii.
15 *Ibid.*, p. xx.
16 Singer, *Giordano Bruno*, p. 363. Singer's book includes a full translation of Bruno's *On the Infinite Universe and Worlds*.
17 *Ibid.*, pp. 322–3.
18 Copenhaver, p. 83.
19 Gingerich, p. 23.
20 Stephen Johnston, 'Like Father, Like Son? John Dee, Thomas Digges and the Identity of the Mathematician', in Clucas (ed.), p. 65.
21 See Westman and McGuire, p. 24.
22 Singer, *Giordano Bruno*, p. 285.
23 Tompkins, p. 83.
24 Gribbin, p. 3.
25 Gatti, *Giordano Bruno and Renaissance Science*, pp. 80–5.
26 Gatti, 'Giordano Bruno's Copernican Diagrams', pp. 43–6.
27 Debus, 'Robert Fludd and the Circulation of the Blood'.
28 *Ibid.*

29 Copenhaver, p. 33.
30 Atanasijevic, p. xvii.
31 *Ibid.*, p. xviii.
32 Yates, *Giordano Bruno and the Hermetic Tradition*, p. 304.
30 Quoted in Tompkins, p. 23.
34 Quoted in Yates, *Giordano Bruno and the Hermetic Tradition*, p. 312.
35 Quoted in *ibid.*, p. 312.
36 See *ibid.*, pp. 320–1.
37 *Ibid.*, p. 341.
38 This is the description given to the extract from Boccalini's work that was included with the first of the Rosicrucian manifestos.
39 Findlen, 'A Hungry Mind'.
37 *Ibid.*

Chapter Three

1 Ferris, pp. 85–6.
2 Yates, *Giordano Bruno and the Hermetic Tradition*, p. 360.
3 *Ibid.*, p. 363.
4 Mason, p. 462.
5 *Ibid.*, p. 468.
6 See Morley for a translation of *City of the Sun*.
7 Interviewed in Burstein and de Keijzer, p. 242.
8 Yates, *Giordano Bruno and the Hermetic Tradition*, p. 233.
9 Quoted in Olaf Pedersen, 'Galileo's Religion', in Coyne (ed.), p. 75.
10 In his notes to Galileo, Salusbury translation, p. 15.
11 Oxford University science historian Allan Chapman, quoted in Couper and Henbest, p. 154.
12 In his forward to Stillman Drake's translation of Galileo, p. xvii.
13 Pedersen, 'Galileo's Religion', in Coyne (ed.), pp. 80–1.
14 Quoted in *ibid.*, p. 80.
15 Yates, *Giordano Bruno and the Hermetic Tradition*, p. 383.
16 This was in a conversation in 1610 with Martin Hasdale, the librarian at Rudolph II's court, who relayed Kepler's remarks to Galileo in a letter. (Singer, *Giordano Bruno*, p. 189.)
17 Bruno, *The Ash Wednesday Supper*, pp. 122–3.
18 Quoted in Finocchario, p. 88.
19 Pedersen, 'Galileo's Religion', in Coyne (ed.), p. 97.
20 *Ibid.*, p. 92.
21 Finocchiaro, p. 13.
22 Quoted in Pedersen, 'Galileo's Religion', in Coyne (ed.), p. 81.
23 *Ibid.*, p. 97.
24 Quoted in *ibid.*, p. 81.
25 Yates, *Giordano Bruno and the Hermetic Tradition*, p. 361.

Chapter Four

1 Fowden, p. xxii.
2 Yates, *Giordano Bruno and the Hermetic Tradition*, p. 21.
3 From Thomas Vaughan's 1652 English translation of the *Fama*, reproduced in the appendix to Yates, *The Rosicrucian Enlightenment*, p. 238.
4 See Churton, *The Golden Builders*, pp. 105–17.
5 Yates, *The Rosicrucian Enlightenment*, p. 250.
6 Churton, *The Golden Builders*, p. 93.
7 *Ibid.*, p. 132.
8 Yates, *The Rosicrucian Enlightenment*, p. 47.
9 Churton, *The Golden Builders*, p. 131.
10 *Ibid.*, p. 143.
11 Theophrastus Bombastus von Hohenheim (1493–1541) – he adopted the name Paracelcus to show he was greater than Celsus, the Roman author of a classic encyclopaedia of medicine – was a Swiss botanist, herbalist and physician. He was heavily influenced by the works of Pico and Ficino, applying the principles of Hermeticism and talismanic magic to healing. His ideas about the combination and manipulation of the elements also led to him to alchemy. Some think that Christian Rosenkreutz was intended to represent Paracelsus, despite the fact that the *Fama* explicitly says that he wasn't a member of the Rosicrucian fraternity, although adding that it did allow him access to the book containing their accumulated wisdom, the 'Book M'.
12 Churton, *The Golden Builders*, p. 157.
13 See Yates, *The Art of Memory*, chapters XV and XVI.
14 Quoted in Yates, *The Rosicrucian Enlightenment*, pp. 101–2.
15 *Ibid.*, p. 136.
16 Purver, p. 223.
17 Quoted in Tompkins, p. 86.
18 Quoted in Yates, *Giordano Bruno and the Hermetic Tradition*, p. 445. (Our translation from the French.)
19 Yates, *The Rosicrucian Enlightenment*, p. 113.

Chapter Five

1 Couturat, p. 131.
2 Yates, *The Art of Memory*, pp. 387–8.
3 *Ibid.*, p. 382.
4 Quoted in the online Stanford Encyclopaedia of Philosophy: plato.stanford.edu/entries/leibniz.
5 Standford Encyclopaedia of Philosophy website: plato.stanford.edu/entries/cambridge-platonists.
6 See Yates, *The Art of Memory*, p. 388, and Atanasijevic, p. xviii.

7 Yates, *The Art of Memory*, p. 388.
8 Quoted in *ibid.*, p. 385.
9 Quoted in *ibid.*
10 Strange Science website: www.strangescience.net/kircher.htm.
11 Quoted in Tompkins, p. 90.
12 *Ibid.*, p. 97.
13 Interviewed in Burstein and de Keijzer, pp. 239–40.
14 See 'Bernini's Elephant and Obelisk' in Hecksher. This is a reproduction of an article that appeared in *The Art Bulletin* in 1947.
15 Quoted in Tompkins, p. 88.
16 Tod Marder, 'A Bernini Expert Reflects on Dan Brown's Use of the Baroque Master', in Burstein and de Keijzer, p. 255.
17 Tompkins, p. 97.
18 Quoted in Ingrid D. Rowland, 'Athanasius Kircher, Giordano Bruno, and the *Panspermia* of the Infinite Universe', in Findlen (ed.), *Athanasius Kircher*, p. 56.
19 See Picknett, *Mary Magdalene*, pp. 27–9.
20 Tompkins, p. 100.
21 Ingrid D. Rowland, 'Athanasius Kircher, Giordano Bruno, and the *Panspermia* of the Infinite Universe', in Findlen (ed.), *Athanasius Kircher*, pp. 201–2.

Chapter Six
1 Quoted in Yates, *The Rosicrucian Enlightenment*, p. 186.
2 Stanford Encyclopaedia of Philosophy, online: plato.stanford.edu/entries/cambridge-platonists.
3 Quoted in Dobbs, *The Foundations of Newton's Alchemy*, p. 115.
4 P. M. Rattansi, 'Some Evaluations of Reason in Sixteenth- and Seventeenth-Century Natural Philosophy', in Teich and Young (eds.), p. 151.
5 Quoted in Yates, *Giordano Bruno and the Hermetic Tradition*, p. 424.
6 Purver, p. 217.
7 Quoted in *ibid.*, pp. 221–2.
8 Quoted in *ibid.*, p. 219.
9 Quoted in *ibid.*, p. 198.
10 Quoted in *ibid.*, p. 199.
11 Bacon, p. 67.
12 Rossi, pp. 13–14.
13 Tuveson, p. 52.
14 Bacon, p. ix.
15 J. R. Ravetz, 'Francis Bacon and the Reform of Philosophy', in Debus (ed.), *Science, Medicine and Society in the Renaissance*, vol. II, p. 101.
16 Bacon, p. 1.

17 *Ibid.*, pp. 2–3.
18 *Ibid.*, p. 3.
19 E.g. Tuveson, pp. 170–9, Yates, *The Rosicrucian Enlightenment*, chapter XV.
20 Lomas, p. 320.
21 From Lomas' lecture 'Sir Robert Moray – Soldier, Scientists, Spy, Freemason and Founder of the Royal Society', given at Gresham College, 4 April 2007. A transcript is available on the Gresham College website: www.gresham.ac.uk/event.aspPageId=45& EventId=589.
22 Quoted in Purver, p. 221.
23 Quoted in *ibid.*, pp. 221–2.
24 Quoted in *ibid.*, p. 232.
25 Quoted in *ibid.*
26 Quoted in Bluhm, p. 185.
27 *Ibid.*, pp. 183–6.
28 Gribbin, p. 229.
29 Lord Rees, today's President of the Royal Society, quoted in Bragg, p. 22.
30 Gribbin, pp. 238–9.
31 Hollis, p. 262.
32 Richard S. Westfall, 'Newton and the Hermetic Tradition', in Debus (ed.), *Science, Medicine and Society in the Renaissance*, vol. II, pp. 185–6.
33 'Newton, the Man' in Keynes, p. 363.
34 *Ibid.*, p. 366.
35 Quoted in Yates, *The Rosicrucian Enlightenment*, p. 200.
36 McGuire and Rattansi, p. 109.
37 *Ibid.*, p. 127.
38 *Ibid.*, p. 124.
39 Westfall, 'Newton and the Hermetic Tradition', in Debus (ed.), *Science, Medicine and Society in the Renaissance*, vol. II, p. 193.
40 Dobbs, *The Janus Face of Genius*, p. 68.
41 *Ibid.*, pp. 185–6.
42 Quoted in Westfall, *Never at Rest*, p. 434.
43 Westfall, 'Newton and the Hermetic Tradition', in Debus (ed.), vol. II, pp. 194–5.
44 Hitchens, p. 65.

Chapter Seven
1 Fowden, pp. 68–74.
2 See below, p. 185.
3 Festugière, p. 102.
4 Luckert, p. 55.

5 Lurker, p. 121.
6 *Ibid.*
7 Ray, p. 65.
8 *Ibid.*, p. 160.
9 Fowden, p. 34.
10 Lurker, pp. 69–70.
11 Ray, p. 165.
12 Fowden, p. 27.
13 *Ibid.*, pp. 40–1.
14 According to Plutarch (p. 161) the establishing of the Serapis cult was the work of Manetho and a member of the family that held the hereditary priesthood of the Greek mystery centre of Eleusis, which makes sense if it was to be a 'hybrid' cult for Egyptians and Greeks. Although some doubt Plutarch's story, Manetho was certainly associated with the cult – see J. Gwyn Griffith's notes to *ibid.*, pp. 387–8.
15 Iamblichus, p. 5.
16 Fowden, p. xxv.
17 Churton, *The Gnostic Philosophy*, p. 120.
18 Plotinus, p. 9.
19 Luckert, p. 261.
20 *Ibid.*, p. 262.
21 Quoted in *ibid.*, p. 260.
22 See *ibid.*, chapter 14.
23 *Ibid.*, p. 257.
24 Eunapius, 'Lives of the Philosophers', in Philostratus and Eunapius, pp. 419–25.
25 Herodotus, p. 130.
26 Luckert, p. 42.
27 E.g. Lurker, p. 99.
28 See Luckert, chapter 2.
29 *Ibid.*, p. 52.
30 Lurker, p. 31.
31 Lehner, p. 34.
32 Luckert, p. 52.
33 *Ibid.*, p. 45.
34 *Ibid.*, p. 57.
35 Campbell and Musès, p. 138.

Chapter Eight

1 'Humanism' is a fluid term, coined in the mid-nineteenth century and applied not just to contemporary ideas but also retrospectively to earlier philosophers and social reformers. It is applied to any

philosophy that places human beings at the centre of things, asserting not only their fundamental right to control their own destiny but also stressing their *ability* to do so. But beyond that, the precise definition varies depending on the era in question: the values and ideals of a twenty-first century humanist are very different from a fifteenth-century one. The biggest difference is that today's humanism tends to eschew the metaphysical and religious. Under this definition, the likes of Pico, Ficino and Bruno qualify as humanists, but they would never have recognized the term.

2 Magee, p. 7.
3 P. M. Rattansi, 'Some Evaluations of Reason in Sixteenth- and Seventeenth-Century Natural Philosophy', in Teich and Young (eds.), p. 149.

Chapter Nine

1 In the radio programme 'The Multiverse', part of the *In Our Time* series, broadcast on BBC Radio 4 on 21 February 2008.
2 Barrow and Tipler, p. 5.
3 Susskind, 'A Universe Like No Other', p. 38.
4 Weinberg, *The First Three Minutes*, p. 154.
5 Carr and Rees, p. 612.
6 Dyson, p. 44.
7 Quoted in Davies, *The Mind of God*, p. 199.
8 Stockwood (ed.), p. 64.
9 Davies, *The Mind of God*, Chapter 8.
10 Feynman, p. 12.
11 Davies, *The Mind of God*, p. 197.
12 In the BBC Radio 4 programme 'The Multiverse' (see note 1 above).
13 Hawking and Mlodinow, p. 161.
14 Davies, *The Goldilocks Enigma*, pp. 166–70.
15 Susskind, 'A Universe Like No Other', p. 37.
16 *Ibid.*, p. 39.
17 In the BBC Radio 4 programme 'The Multiverse' (see note 1 above).
18 Smolin, *The Trouble with Physics*, pp. 166–7.
19 Jeans, p. 96.
20 Davies, *The Mind of God*, p. 173.
21 In the BBC Radio 4 programme 'The Multiverse' (see note 1 above).
22 Carr, p. 14.
23 *Ibid.*
24 Quoted in Smolin, *The Trouble With Physics*, p. 125.
25 *Ibid.*, pp. 158–9.
26 Al-Khalili, p. 23.
27 Smolin, *The Trouble with Physics*, p. 163.
28 Quoted in Malone, p. 191.

29 See Nick Bostrom, 'Are We Living in *The Matrix*? The Simulation Argument', in Yeffeth (ed.).

30 Davies, *The Goldilocks Enigma*, pp. 213–4.

31 Hawking, 'The Grand Designer', p. 25.

32 Al-Khalili, p. 23.

33 Weinberg, *Dreams of a Final Theory*, p. 182.

34 Susskind, 'A Universe Like No Other', p. 36.

35 Weinberg, *Dreams of a Final Theory*, p 182.

36 Hoyle, pp. 217–8.

37 Davies, *The Mind of God*, p. 16.

38 Quoted in Schönborn.

39 *Ibid*.

40 Davies, *The Goldilocks Enigma*, pp. 228–30.

Chapter Ten

 1 Quoted in Dawkins, *The Greatest Show on Earth*, p. 417.

 2 *Ibid.*, p. 416.

 3 De Duve, *Vital Dust*, p. xv.

 4 Watson and Crick, p. 738.

 5 See Ingrid D. Rowland, 'Athanasius Kircher, Giordano Bruno, and the *Panspermia* of the Infinite Universe', in Findlen (ed.).

 6 Hoyle and Wickramasinghe, *Evolution from Space*, p. xiii–xv.

 7 Quoted in Carey.

 8 Quoted on BBC News website, ' "Life Chemical" Detected in Comet', 18 August 2009: news.bbc.co.uk/1/hi/sci/tech/8208307.stm.

 9 Schueller, p. 34.

10 Quoted in *ibid.*, p. 31.

11 *Ibid.*, p. 34.

12 Quoted in *ibid.*, p. 35

13 Lovelock, *Gaia* (1979 edition), p. vii.

14 In the documentary, 'Life, the Universe and *Everything*: James Lovelock' in the *Beautiful Minds* series, produced and directed by Paul Bernays, ARC Productions for BBC Four, 2010.

15 Interviewed in the above documentary.

16 Lovelock, *Gaia* (2000 edition), p. xv.

17 *Ibid.*, p. ix.

18 Lovelock, *The Ages of Gaia*, p. 15.

19 De Duve, *Vital Dust*, p. 20.

20 *Ibid*, pp. 286–9.

21 *Ibid.*, pp. 292–3.

Chapter Eleven

1 E.g. Dawkins, *The God Delusion*, p. 173.
2 Crick, p. 58.
3 Monod, p. 167.
4 Hoyle and Wickramasinghe, *Evolution from Space*, p. 119.
5 Davies, *The Cosmic Blueprint*, p. 109.
6 Smith, *Did Darwin Get It Right?*, p. 167.
7 Crick, p. 113.
8 Narby, *The Cosmic Serpent*, p. 92.
9 De Duve, *Life Evolving*, p. 51.
10 See Leipe, Aravina and Koonin.
11 Hamilton, p. 29.
12 *Ibid.*
13 In his Gifford Lecture 'Life's Solution: The Predictability of Evolution Across the Galaxy (and Beyond)', given at the University of Edinburgh on 19 Feb 2007. Audio file available at the University of Edinburgh's Humanities and Social Science's website: www.hss.ed.ac.uk/giffordexemp/2000/details/ProfessorSimonCo nwayMorris.html.
14 Dawkins, *The God Delusion*, pp. 164–5.
15 Cavalier-Smith, p. 998.
16 Prokaryotes have, since Carl Woese's discovery in 1977, been divided between bacteria and archaea, as described above, but neither this nor the evolution of the apparent independent DNA of bacteria, affects our point here.
17 Cavalier-Smith, p. 978.
18 *Ibid.*
19 Margulis and Sagan, pp. 115–6.
20 *Ibid.*, p. 118.
21 Quoted in Ridley, p. 315.
22 Williams, p. v.
23 *Ibid.*, p. 11.
24 Smith, *The Evolution of Sex*, p. 10.
25 Smith, *Did Darwin Get It Right?*, p. 165.
26 Ridley, p. xxii.
27 Smith, *Did Darwin Get It Right?*, p. 165.
28 Williams, p. 8.
29 Margulis and Sagan, p. 157.
30 Smith, *Did Darwin Get It Right?*, pp. 166–7.
31 Williams, p. 11.
32 See Guarente and Kenyon.
33 A. M. Leroi, A. K. Chippindale and M. R. Rose, 'Long-Term Laboratory Evolution of a Genetic Life-History Trade-Off in *Drosophila Melanogaster*', in Rose, Passananti and Matos (eds.). (This

is a reproduction of a paper that first appeared in the journal *Evolution* in 1994.)

34 Stephen Jay Gould, 'G. G. Simpson, Paleontology, and the Modern Synthesis', in Mayr and Provine, pp. 153–4.

35 Mayr, pp. 529–30. A genus is the next step up from a species in biological classification, a group of distinct species that are closely related genetically, sharing a close common ancestor. Examples are the genera *Canis*, to which dogs, wolves, jackals, coyotes and dingoes belong, and *Equus*, which includes horses, donkeys and zebras.

36 In the radio show 'The Whale – A History', in the *In Our Time* series presented by Melvyn Bragg, broadcast on BBC Radio 4 on 21 May 2009.

37 See, for example, Smith, *Did Darwin Get It Right?*, pp. 148–9.

38 Goodwin, pp. xii–xiii.

39 Many have the impression from the title of his book *The Selfish Gene* that Richard Dawkins proposes that natural selection acts at the level of the gene. But he doesn't: he argues that evolution should be *viewed* from the level of genes, because animals and plants are basically big bags of genes. Natural selection acts on the individual, but its ultimate effect is on the gene pool of the species, determining what genes are in it and how many of each gene there are. Although offering a potentially useful new perspective for evolutionists to look at certain questions, this theory ultimately only describes the same things in different words.

40 Fort, p. 38.

41 Le Page, p. 26.

42 Dawkins, *The Greatest Show on Earth*, pp. 297–8.

43 See Dawkins, *Climbing Mount Improbable*, chapter 5.

44 Mayr, p. 541.

45 Popper, p. 171.

46 *Ibid.*, p. 168.

47 *Ibid.*, p. 172.

48 Smith, *Did Darwin Get It Right?*, p. 180.

49 Smith, *The Evolution of Sex*, p. ix.

50 Dawkins, *The Blind Watchmaker*, p. 287.

51 Conway Morris, *Life's Solution*, pp. 315–6.

52 See Mayr's preface to Mayr and Provine, pp. ix–x.

53 Mayr and Provine, p. xv.

54 Stephen Jay Gould, 'The Hardening of the Modern Synthesis', in Grene (ed.), p. 88.

55 *Ibid.*, p. 90.

56 *Ibid.*, p. 91.

57 Dawkins, *The Ancestor's Tale*, p. 262.

58 Quoted in Costas R. Krimbas, 'The Evolutionary Worldview of Theodosius Dobzhansky', in Adams (ed.), p. 188.
59 Dobzhansky, *Genetics of the Evolutionary Process*, p. 430.
60 *Ibid.*, p. 431.
61 Costas R. Krimbas, 'The Evolutionary Worldview of Theodosius Dobzhansky', in Adams (ed.), p. 189.
62 Dobzhansky, *Genetics of the Evolutionary Process,* p. 391.
63 Teilhard de Chardin, p. 258.
64 See Curtis L. Hancock, 'The Influence of Plotinus on Bergson's Critique of Empirical Science', in Harris, vol. I.
65 Bergson, p. 384.
66 Barrow and Tipler, p. 204.

Chapter Twelve
1 Popper, p. 173.
2 Conway Morris, *Life's Solution*, p. 316.
3 In his sixth and final Gifford Lecture, 'Towards an Eschatology of Evolution', at the University of Edinburgh, 1 March 2007. Audio file available at the University of Edinburgh's Humanities and Social Science's website: www.hss.ed.ac.uk/giffordexemp/2000/details/ProfessorSimonConwayMorris.
4 Conway Morris, *Life's Solution*, p. xv.
5 *Ibid.*, p. xii.
6 Abstract to Conway Morris' Gifford Lecture 'Life's Solution', see Chapter 11, note 13.
7 Conway Morris, *Life's Solution*, pp. 292–5.
8 In his fourth Gifford Lecture 'Becoming Human: The Continuing Mystery', given at the University of Edinburgh on 26 Feb 2007. Audio file available at the University of Edinburgh's Humanities and Social Science's website: www.hss.ed.ac.uk/giffordexemp/2000/details/ProfessorSimonConwayMorris.
9 Conway Morris, *Life's Solution*, p. 328.
10 Shapiro, p. 807.
11 De Duve, *Vital Dust*, p. 297.
12 In his fourth Gifford Lecture – see note 8 above.
13 Polanyi, p. 47.
14 From the abstract of his fourth Gifford Lecture – see note 8 above.
15 'Who Speaks for the Earth?', thirteenth and final episode of the TV series *Cosmos*, first broadcast 21 December 1980. DVD released by Freemantle Home Entertainment, 2009. Directed by David F. Oyster, written by Carl Sagan, Ann Druyan and Steven Soter.
16 In the documentary movie *A Brief History of Time,* produced by David Hickman and directed by Errol Morris, Anglia Television/Gordon Freedman Productions, 1991.

17 The papers were published in Halliwell, Pérez-Mercader and Zurek.
18 Bierman, 'A World With Retroactive Causation', p. 1.
19 Davies, *The Goldilocks Enigma*, p. 274.
20 George, p. 56.
21 *Ibid*.
22 Bierman and Houtkooper.
23 See Bierman, 'Exploring Correlations Between Local Emotional and Global Emotional Events and the Behavior of a Random Number Generator'.
24 Hagel and Tschapke.
25 Radin, 'Exploring Relationships Between Random Physical Events and Mass Human Attention', p. 538
26 Radin, *Entangled Minds*, p. 206.
27 Wheeler, *Geons, Black Holes, and Quantum Foam*, p. 334.
28 Quoted in Jacques et al, p.1.
29 Interviewed for 'The Anthropic Universe', *The Science Show*, ABC National Radio, 18 February 2006, presented by Martin Redfern, produced by Pauline Newman. Transcript available at: www.abc.net.au/rn/scienceshow/stories/2006/1572643.
30 Wheeler, *Geons, Black Holes, and Quantum Foam*, p. 331.
31 *Ibid*., p. 333.
32 Gardner and Wheeler.
33 Jacques *et al*.
34 Wheeler, *Geons, Black Holes, and Quantum Foam*, p. 337.
35 Davies and Gribbin, p. 208.
36 Wheeler, from his foreword to Barrow and Tipler, p. 6.
37 John Archibald Wheeler, 'Law Without Law', in Wheeler and Zurek (eds.), p. 194.
38 On *The Science Show*, ABC National Radio. See note 29 above.
39 John Archibald Wheeler, 'Genesis and Observership', in Butts and Hintikka (eds.), p. 3.
40 B. J. Carr, 'On the Origin, Evolution and Purpose of the Physical Universe', in Leslie (ed.), p. 152.
41 John Archibald Wheeler, 'Genesis and Observership', in Butts and Hintikka (eds.), p. 21.
42 *Ibid*., p. 19.
43 Barrow and Tipler. p. 203.
44 John Archibald Wheeler, 'Beyond the Edge of Time', in Leslie (ed.), p. 214.
45 Hawking and Mlodinow, p. 140
46 Gefter, p. 30.
47 Barrow and Tipler, p. 470.

48 On *The Science Show*, ABC National Radio. See note 28 above.

49 P. M. Rattansi, 'Newton's Alchemical Studies', in Debus (ed.) *Science, Medicine and Society in the Renaissance*, p. 179.

50 Hawking, *A Brief History of Time*, p. 175.

51 Copenhaver, p. 41.

52 *Ibid.*, p. 37.

53 Magee, p. 10.

54 *Ibid.*

55 Jantsch, p. 308.

56 *Ibid.*, pp. 308–9.

57 Wheeler, 'Law without Law', in Wheeler and Zurek, p. 209.

Chapter Thirteen

1 Richard S. Westfall, 'Newton and the Hermetic Tradition', in Debus (ed.), *Science, Medicine and Society in the Renaissance*, p. 195.

2 John Archibald Wheeler, 'Beyond the End of Time', in Leslie (ed.), p. 212.

3 Richard S. Westfall, 'Newton and the Hermetic Tradition', in Debus (ed.), *Science, Medicine and Society in the Renaissance*, p. 185.

4 David Fideler, 'Neoplatonism and the Cosmological Revolution: Holism, Fractal Geometry, and Mind-in-Nature', in Harris (ed.), vol. I, p. 104.

5 *Ibid.*, p. 106.

6 *Ibid.*, p. 117.

7 Luckert, p. 61.

8 National Constitution Centre website: www.constitutioncenter. org/libertymedal/recipient_1994_speech.

9 *Ibid.*

10 *Ibid.*

11 *Ibid.*

12 Copenhaver, p. 65. (Treatise XVIII)

13 *Ibid.*, p. 36. (Treatise X)

14 Quoted in Fideler, 'Neoplatonism and the Cosmological Revolution: Holism, Fractal Geometry, and Mind-in-Nature', in Harris (ed.), vol. I, p. 116.

15 Copenhaver, p. 48 (Treatise XI).

16 'What We'll Never Know', Rees' third Reith lecture, broadcast on BBC Radio 4 on 16 June 2010. A transcript is available at: downloads.bbc.co.uk/rmhttp/radio4/transcripts/20100615–reith.rtf.

Appendix

1 Copenhaver, p. xliv.

2 Fowden, p. 4.

3 Jonas, Chapter Seven.

4 See Yamauchi, Chapter one.

5 There is controversy over whether the few surviving writings ascribed to Simon Magus – which we only have because they were quoted by early Christian writers as fodder for hellfire and damnation fulmination – were written by him or his followers, but either way they reflect his theology and philosophy.

6 Quoted in Luckert, p. 301.

7 *Ibid.*, pp. 299–308.

8 See *The Masks of Christ*, p. 243–51.

9 G. Quispel, 'The *Asclepius* – From the Hermetic Lodge in Alexandria to the Greek Eucharist and the Roman Mass', in van den Broek and Hanagraff, p. 75.

10 Picknett and Prince, *The Masks of Christ*, pp. 222–5.

SELECT BIBLIOGRAPHY

Entries are for the editions cited. Where this is not the first edition, details of the original publication follow.

Adams, Mark B., ed., *The Evolution of Theodosius Dobzhansky: Essays on His Life and Thought in Russia and America*, Princeton University Press, Princeton, 1994.

Agrippa of Nettesheim, Henry Cornelius, (trans. James Freake, ed. Donald Tyson), *Three Books of Occult Philosophy*, Llwellyn Publications, St. Paul, 1993.

Al-Khalili, Jim, 'M Stands for Maybe', *Eureka*, no. 12, September 2010.

Atanasijevi, Ksenija, *The Metaphysical and Geometrical Doctrine of Bruno as Given in His Work* De Triplici Minimo, Warren H. Green, St Louis, 1972 (*La doctrine métaphyisque et géométrique de Bruno, exposée dans son ouvrage 'De triplici minimo'*, Les Presses Universitaires de France, Belgrade, 1923).

Bacon, Francis, (ed. G. W. Kitchin), *The Advancement of Learning*, J.M. Dent & Sons, London, 1973.

Balcombe, Jonathan, *Second Nature: The Inner Lives of Animals*, Palgrave Macmillan, London, 2010.

Barrow, John D., *New Theories of Everything: The Quest for Ultimate Explanation*, Oxford University Press, Oxford, 2007 (*Theories of Everything*, Oxford University Press, Oxford, 1991).

Barrow, John D., and Frank J. Tipler, *The Anthropic Cosmological Principle*, revised edition, Oxford University Press, Oxford, 1988 (first edition 1986).

Bergson, Henri, *Creative Evolution*, The Modern Library, New York, 1944 (*L'évolution créatrice*, Felix Alcan, Paris, 1907).

Bierman, Dick J., 'A World with Retroactive Causation', University of

Amsterdam website: www.uva.nl/publications/1987/imposworlds 87.pdf.

— Bierman, Dick J., 'Exploring Correlations Between Local Emotional and Global Emotional Events and the Behavior of a Random Number Generator', *Journal of Scientific Exploration*, vol. 10, no. 3, 1996.

Bierman, D. J., and J. M. Houtkooper, 'Exploratory PK Tests with a Programmable High Speed Random Number Generator', *European Journal of Parapsychology*, vol. 1, no. 1, 1975.

Bluhm, R. K., 'Henry Oldenburg, F. R. S. (*c*.1615–1677)', *Notes and Records of the Royal Society of London*, vol. 15, July 1960.

Bostrom, Nick, *Anthropic Bias: Observation Selection Effects in Science and Philosophy*, Routledge, London, 2002.

Bragg, Melvin, *Twelve Books that Changed the World*, Hodder & Stoughton, London, 2006.

Brown, Dan, *Angels and Demons*, Pocket Books, New York, 2000.

Bruno, Giordano, (trans. and ed. Paul Eugene Memmo, Jr.), *The Heroic Frenzies*, University of North Carolina Press, Chapel Hill, 1964.

— (trans. and ed. Arthur D. Imerti), *The Expulsion of the Triumphant Beast*, Rutgers University Press, New Brunswick, 1964.

— (trans. and ed. Stanley L. Jaki), *The Ash Wednesday Supper: La Cena de le Ceneri*, Mouton, The Hague, 1975.

Burstein, Dan, and Arne de Keijzer, *Secrets of Angels & Demons: The Unauthorised Guide to the Bestselling Novel*, Weidenfeld & Nicolson, London, 2005 (CDS Books, New York, 2004).

Butts, Robert E., and Jaakko Hintikka, (eds.), *Foundational Problems in the Special Sciences: Part Two of the Proceedings of the Fifth International Conference of Logic, Methodology and Philosophy of Science, London, Ontario, Canada, 1975*, D. Reidel, Dordrecht, 1977.

Cairns-Smith, A. G., *Seven Clues to the Origins of Life: A Scientific Detective Story*, Cambridge University Press, Cambridge, 1985.

Caldwell, Ian, and Dustin Thomason, *The Rule of Four: A Novel*, Dial Press, New York, 2004.

Campbell, Joseph, and Charles Musès, *In All Her Names: Exploration of the Feminine in Divinity*, Harper SanFrancisco, San Francisco, 1991.

Carey, Bjorn, 'Life's Building Blocks "Abundant in Space"', Space.com website (18 October 2005): www.space.com/scienceastronomy/051018 _science_tuesday.

Carr, Bernard, (ed.), *Universe or Multiverse?*, Cambridge University Press, Cambridge, 2007.

Carr, B. J., and M. J. Rees, 'The Anthropic Principle and the Structure of the Physical World', *Nature*, vol. 278, 12 April 1979.

Cavalier-Smith, Thomas, 'Cell Evolution and Earth History: Stasis and Revolution', *Philosophical Transactions of the Royal Society B*, vol. 361, no. 1470, June 2006.

Churton, Tobias, *The Golden Builders: Alchemists, Rosicrucians and the First Free Masons*, Signal Publishing, Lichfield, 2002.
— *The Gnostic Philosophy*, Signal Publishing, Lichfield, 2003.
Clucas, Stephen, (ed.), *John Dee: Interdisciplinary Studies in English Renaissance Thought*, Springer, Dordrecht, 2006.
Conway Morris, Simon, *Life's Solution: Inevitable Humans in a Lonely Universe*, Cambridge University Press, Cambridge, 2003.
Copenhaver, Brian P., *Hermetica: The Greek* Corpus Hermeticum *and the Latin* Aclepius *in a New English Translation*, Cambridge University Press, Cambridge, 1992.
Couper, Heather, and Nigel Henbest, *The History of Astronomy*, Cassell Illustrated, London, 2007.
Couturat, Louis, *La logique de Leibniz d'après des documents inédits*, Felix Alcan, Paris, 1901.
Coyne, C. V., (ed.), *The Galileo Affair: A Meeting of Science and Faith – Proceedings of the Cracow Conference, May 24–27, 1984*, Specola Vaticana, Vatican City, 1985.
Crick, Francis, *Life Itself: Its Origin and Nature*, Futura, London, 1982 (Simon & Schuster, New York, 1981).
Darwin, Charles, *The Origin of Species by Means of Natural Selection, or the Preservation of Favoured Races in the Struggle for Life*, sixth edition, John Murray, London, 1872 (*On the Origin of Species by Means of Natural Selection*, John Murray, London, 1859).
— *The Descent of Man, and Selection in Relation to Sex*, 2 vols., John Murray, London, 1871.
Davies, Paul, *God and the New Physics*, J. M. Dent & Sons, London, 1983.
— *The Cosmic Blueprint: New Discoveries in Nature's Creative Ability to Order the Universe*, Templeton Foundation Press, Philadelphia, 2004 (Simon & Schuster, New York, 1988).
— *The Mind of God: Science and the Search for Ultimate Meaning*, Penguin, London, 1993 (Simon & Schuster, New York, 1992).
— *The Goldilocks Enigma: Why is the Universe Just Right for Life?*, Allen Lane, London, 2006.
Davies, Paul, and John Gribbin, *The Matter Myth: Towards 21st-Century Science*, Viking, London, 1991.
Dawkins, Richard, *The Extended Phenotype: The Gene as the Unit of Selection*, Freeman, Oxford, 1982.
— *The Blind Watchmaker*, Longman Scientific and Technical, Harlow, 1986.
— *Climbing Mount Improbable*, Viking, London, 1996.
— (with Yan Wong), *The Ancestor's Tale: A Pilgrimage to the Dawn of Life*, Weidenfeld & Nicolson, London, 2004.
— *The Selfish Gene*, updated edition, Oxford University Press, Oxford, 2006 (first edition 1976).

— *The God Delusion*, updated edition, Black Swan, London, 2007 (Bantam Press, London, 2006).

— *The Greatest Show on Earth: The Evidence for Evolution*, Bantam Press, London, 2009.

Debus, Allen G., 'Robert Fludd and the Civilization of the Blood', Journal of the History of Medicine and Allied Sciences, vol. XVI, no. 4, 1961. *Science, Medicine and Society in the Renaissance: Essays to Honor Walter Pagel*, 2 vols., Science History Publications, New York, 1972.

de Duve, Christian, *Vital Dust: Life as a Cosmic Imperative*, Basic Books, New York, 1995.

— *Life Evolving: Molecules, Mind and Meaning*, Oxford University Press, New York, 2002.

— *Singularities: Landmarks on the Pathways of Life*, Cambridge University Press, New York, 2005.

Dobbs, Betty Jo Teeter, *The Foundations of Newton's Alchemy, or 'The Hunting of the Greene Lyon'*, Cambridge University Press, Cambridge, 1983 (first edition 1975).

— *The Janus Face of Genius: The Role of Alchemy in Newton's Thought*, Cambridge University Press, Cambridge, 1991.

Dobzhansky, Theodosius, *Genetics and the Origin of Species*, revised edition, Columbia University Press, New York, 1951 (first edition 1937).

— *Evolution, Genetics and Man*, Wiley, New York, 1955.

— *Genetics of the Evolutionary Process*, Columbia University Press, New York, 1970.

Dyson, Freeman J., *A Many Colored Glass: Reflections on the Place of Life in the Universe*, University of Virginia Press, Charlottesville, 2007.

Faivre, Antoine, *The Eternal Hermes: From Greek God to Alchemical Magus*, Phanes Press, Grand Rapids, 1995.

Faulkner, R. O., *The Ancient Egyptian Pyramid Texts*, Oxford University Press, Oxford, 1969.

Ferris, Timothy, *Coming of Age in The Milky Way*, The Bodley Head, London, 1989 (William Morrow & Co., New York, 1988).

Festugière, R. P., *La révélation d'Hermès Trismégiste*, 4 vols., J. Gabalda, Paris, 1949–54.

Feynman, Richard P., *The Meaning of It All*, Penguin Books, London, 1998.

Finocchiaro, Maurice A., trans. and ed., *The Galileo Affair: A Documentary History*, University of California Press, Berkeley and Los Angeles, 1989.

Findlen, Paula, *Athanasius Kircher: The Last Man Who Knew Everything*, Routledge, New York/London, 2004.

— 'A Hungry Mind: Giordano Bruno, Philosopher and Heretic', *The Nation*, 29 September 2008.

Fort, Charles, *The Book of the Damned*, Abacus, London, 1974, (Boni and Liveright, New York, 1919).

Fowden, Garth, *The Egyptian Hermes: A Historical Approach to the Late Pagan Mind*, Cambridge University Press, Cambridge, 1986.

Freke, Timothy, and Peter Gandy, *The Hermetica: The Lost Wisdom of the Pharaohs*, Piatkus, London, 1997.

French, Peter, *John Dee: The World of an Elizabethan Magus*, Routledge and Kegan Paul, London, 1972.

Galileo, (trans. Thomas Salusbury, rev. and ed. Giorgio de Santillana), *Dialogue on the Two Great World Systems, in the Salusbury Translation*, University of Chicago Press, Chicago, 1953 (first published 1661).

— (trans. and ed. Stillman Drake), *Dialogue Concerning the Two Chief World Systems – Ptolemaic and Copernican*, revised edition, University of California Press, Berkeley and Los Angeles, 1967 (first edition 1953).

— (trans. and ed. Maurice A. Finocchiaro), *Galileo on the World Systems: An Abridged Translation and Guide*, University of California Press, Berkeley and Los Angeles, 1997.

Gardner, Martin and John Archibald Wheeler, 'Quantum Theory and Quack Theory', *New York Review of Books*, vol. 26, no. 8, 17 May 1979.

Gatti, Hilary, *Giordano Bruno and Renaissance Science*, Cornell University Press, Ithaca, 1999.

— 'Giordano Bruno's Copernican Diagrams, *Filozotski vestnik*, vol. xxv, no. 2, 2004.

Gefter, Amanda, 'Mr Hawking's Flexiverse', *New Scientist*, vol. 190, no. 2548, 22 April 2006.

George, Alison, 'Lone Voices Special: Take Nobody's Word for It', *New Scientist*, no. 2581, 9 December 2006.

Gingerich, Owen, *The Book Nobody Read: Chasing the Revolutions of Nicolaus Copernicus*, Arrow, London, 2004.

Gleick, James, *Isaac Newton*, Fourth Estate, London/New York, 2003.

Godwin, Joscelyn, *Athanasius Kircher: A Renaissance Man and the Quest for Lost Knowledge*, Thames & Hudson, London, 1979.

— *Robert Fludd: Hermetic Philosopher and Surveyor of Two Worlds*, Thames & Hudson, London, 1979.

— *Athanasius Kircher's Theatre of the World*, Thames & Hudson, London, 2009.

Goodwin, Brian, *How the Leopard Changes Its Spots: The Evolution of Complexity*, updated edition, Princeton University Press, Princeton, 2001 (Charles Scriber's Sons, New York, 1994).

Grene, Marjorie, ed., *Dimensions of Darwinism: Themes and Counterthemes in Twentieth-Century Evolutionary Theory*, Cambridge University Press, London/Éditions de la Maison des Sciences de l'Homme, Paris, 1983.

Gribbin, John, *The Fellowship: The Story of a Revolution*, Allen Lane, London, 2005.

Guarente, Leonard, and Cynthia Kenyon, 'Genetic Pathways That Regulate Ageing in Model Organisms', *Nature*, vol. 408, no. 6809, 9 November 2000.

Gündüz, Sinasi, *The Knowledge of Life: The Origins and Early History of the Mandaeans and Their Relation to the Sabians of the Qur'an and to the Harranians*, Oxford University Press, Oxford, 1994.

Hagel, Johannes, and Margot Tschapke, 'Setup for an Exploratory Study of Correlations Between Collective Emotional Events and Random Number Sequences', paper presented at the Parapsychological Association Convention, University of Vienna, available at the Parapsychological Association website: www.parapsych.org/papers/40.pdf, August 2004.

Halliwell, J. J., J. Pérez-Mercader and W.H. Zurek, *Physical Origins of Time Asymmetry*, Cambridge University Press, Cambridge, 1994.

Hamilton, Garry, 'Looking for LUCA – the Mother of All Life', *New Scientist*, no. 2515, 3 September 2005.

Harris, R. Baines, (ed.), *Neoplatonism and Contemporary Thought*, 2 vols., State University of New York Press, Albany, 2002.

Hawking, Stephen W., *A Brief History of Time: From the Big Bang to Black Holes*, Bantam Press, London, 1988.

— (ed.), *On the Shoulders of Giants: The Great Works of Physics and Astronomy*, Running Press, Philadelphia, 2002.

— 'The Grand Designer', *Eureka*, no. 12, September 2010.

Hawking, Stephen, and Leonard Mlodinow, *The Grand Design*, Bantam Press, London, 2010.

Hawking, Stephen, and Roger Penrose, *The Nature of Space and Time*, Princeton University Press, Princeton, 1996.

Hecksher, William S., (ed. Egon Verheyen), *Art and Literature: Studies in Relationship*, Verlag Valentin Koerner, Baden-Baden, 1985.

Herodotus, (trans. Aubrey de Sélincourt, rev. and ed. by A. R. Burn), *The Histories*, Penguin, London, 1972 (first edition 1954).

Hitchens, Christopher, *God is Not Great: The Case Against Religion*, Atlantic Books, New York, 2007.

Hollis, Leo, *The Phoenix: St Paul's Cathedral and the Men Who Made Modern London*, Weidenfeld and Nicolson, London, 2008.

Holmes, Bob, 'Second Genesis', *New Scientist*, vol. 201, no. 2699, 14 March 2009.

Hoyle, Fred, *The Intelligent Universe*, Michael Joseph, London, 1983.

Hoyle, Fred, and Chandra Wickramasinghe, *Lifecloud: The Origins of Life in the Universe*, J.M. Dent & Sons, London, 1978.

— *Evolution from Space*, Granada, London, 1983 (J. M. Dent & Sons, London, 1981).

— *Proofs That Life is Cosmic*, Institute for Fundamental Studies, Colombo, 1982.
— *From Grains to Bacteria*, University College Cardiff, Cardiff, 1984.
— *Cosmic Life-Force*, J. M. Dent & Sons, London, 1988.
— *Our Place in the Cosmos: The Unfinished Revolution*, J. M. Dent & Sons, London, 1993.
— (eds.), *Astronomical Origins of Life: Steps Towards Panspermia*, Kluwer Academic Publishers, Dordrecht, 2000.
Hudgins, Douglas M., Charles W. Bauschlicher, Jr., and L. J. Allamandola, 'Variations in the Peak Position of the 6.2 m Interstellar Emission Feature: A Tracer of N in the Interstellar Polycyclic Aromatic Hydrocarbon Population', *Astrophysical Journal*, vol. 632, no. 1, October 2005.
Iamblichus, (trans. Emma C. Clarke, John M. Dillon and Jackson P. Hershbell), *De mysteriis*, Brill, Leiden, 2004.
Iliffe, Rob, *Newton: A Very Short Introduction*, Oxford University Press, Oxford, 2007.
Jacques, V., E. Wu, F. Grosshans, F. Treussart, P. Grainger, A. Aspect and J-F. Roch, 'Experimental Realization of Wheeler's Delayed-Choice Gedanken Experiment', *Science*, 315, 5814, 2007.
Jahn, Robert G., *The Role of Consciousness in the Physical World*, Westview Press, Boulder, 1981.
Jantsch, Erich, *The Self-Organizing Universe: Scientific and Human Implications of the Emerging Paradigm of Evolution*, Pergamon Press, Oxford, 1980.
Jonas, Hans, *The Gnostic Religion: The Message of the Alien God and the Beginnings of Christianity*, revised edition, Routledge, London, 1992 (first edition 1958).
Jeans, Sir James, *The Mysterious Universe*, University Press, Cambridge, 1930.
Kepler, Johannes, (trans. and ed. E. J. Aiton, A. M. Duncan and J. V. Field), *The Harmony of the World*, American Philosophical Society, Philadelphia, 1997.
Keynes, John Maynard, *The Collected Writings of John Maynard Keynes, Vol X: Essays in Biography*, expanded edition, Macmillan, London, 1972 (first edition 1933).
Kingsford, Anna, and Edward Maitland (trans. and eds.), *The Virgin of the World of Hermes Mercurius Trismegistus*, George Redway, London, 1885.
Leake, Jonathan, and Andrew Sniderman, 'We are Born to Believe in God', *Sunday Times*, 6 September 2009.
Lehner, Mark, *The Complete Pyramids*, Thames and Hudson, London, 1997.
Leibniz, Gottfried Wilhelm, (trans. George MacDonald Ross), *The Monadology*, University of Leeds Department of Philosophy website:

www.philosophy.leeds.ac.uk/GMR/hmp/texts/modern/leibniz/m onadology/monadology.

Leipe, Detlef D., L. Aravind and Eugene V. Koonin, 'Did DNA Replication Evolve Twice Independently?', *Nucleic Acids Research*, vol. 27, no. 17, September 1999.

Le Page, Michael, 'Evolution: A Guide for the Not-yet Perplexed', *New Scientist*, vol. 198, no. 2652, 19 April 2008.

Leslie, John, (ed.), *Physical Cosmology and Philosophy*, Macmillan, New York/Collier Macmillan, London, 1990.

Lindsay, Jack, *The Origins of Alchemy in Graeco-Roman Egypt*, Frederick Muller, London, 1970.

Lomas, Robert, *The Invisible College: The Royal Society, Freemasonry and the Birth of Modern Science*, Headline, London, 2002.

Lovelock, James, *Gaia: A New Look at Life on Earth*, revised edition, Oxford University Press, Oxford, 2000 (first edition 1979).

— *The Ages of Gaia: A Biography of Our Living Earth*, Oxford University Press, Oxford, 1995 (first edition 1988).

Luckert, Karl W., *Egyptian Light and Hebrew Fire: Theological and Philosophical Roots of Christendom in Evolutionary Perspective*, State University of New York Press, Albany, 1991.

Lurker, Manfred, *An Illustrated Dictionary of the Gods and Symbols of Ancient Egypt*, Thames & Hudson, London, 1982 (*Götter und Symbole der Alten Ägypter*, Barth, Munich, 1974).

Magee, Glenn Alexander, *Hegel and the Hermetic Tradition*, Cornell University Press, Ithaca, 2001.

Mahé, Jean-Pierrer, *Hermès en Haute-Egypte*, 2 vols., Les Presses de l'Université Laval, Quebec, 1978/82.

Malone, John, *Unsolved Mysteries of Science: A Mind-Expanding Journey Through a Universe of Big Bangs, Particle Waves and Other Perplexing Concepts* (Wiley, New York, 2001).

Margulis, Lynn, and Dorion Sagan, *Microcosmos: Four Billion Years of Evolution from Our Microbial Ancestors*, University of California Press, Berkeley and Los Angeles, 1997 (Summit Books, New York, 1986).

Mason, Stephen, 'Religious Reformation and the Pulmonary Transit of the Blood', *History of Science*, vol. 41, part 4, no. 134, December 2003.

Mayr, Ernst, *Toward a New Philosophy of Biology: Observations of an Evolutionist*, Belknap Press, Cambridge, 1988.

Mayr, Ernst, and William B. Provine, *The Evolutionary Synthesis: Perspectives on the Unification of Biology*, Harvard University Press, Cambridge, 1998 (first edition 1980).

McGuire, J. E., and P. M. Rattansi, 'Newton and the "Pipes of Pan"', *Notes and Records of the Royal Society of London*, vol. 21, no. 2, December 1966.

Mead, G. R. S., *Thrice-Greatest Hermes: Studies in Hellenistic Theosophy and Gnosis*, 3 vols., Theosophical Publishing Society, London, 1906.

Michel, Paul-Henri, *The Cosmology of Giordano Bruno*, Methuen, London, 1973 (*La cosmologie de Giordano Bruno*, Hermann, Paris, 1962).

Monod, Jacques, *Chance and Necessity: An Essay on the Natural Philosophy of Modern Biology*, Collins, London, 1972 (*Le hasard et la nécessité*, Éditions du Seuil, Paris, 1970).

Morley, Henry, (ed.), *Ideal Commonwealths: Plutarch's* Lycurgus, *More's* Utopia, *Bacon's* New Atlantis, *Campanella's* City of the Sun, *and a Fragment of Hall's* Mundus alter et idem, George Routledge & Sons, London, 1890.

Narby, Jeremy, *The Cosmic Serpent, DNA and the Origins of Knowledge*, Phoenix, London, 1999 (*Le serpent cosmique, l'ADN et les origines du savior*, Georg Editeur, Geneva, 1995).

— *Intelligence in Nature: An Enquiry into Knowledge*, Jeremy P. Tarcher/Penguin, New York, 2005.

Parks, Tim, *Medici Money: Banking, Metaphysics and Art in Fifteenth-Century Florence*, Profile Books, London, 2005.

Penrose, Roger, *The Emperor's New Mind: Concerning Computers, Minds, and the Laws of Physics*, Oxford University Press, Oxford, 1999 (first edition 1989).

Philostratus and Eunapius, (trans. Wilmer Cave Wright), *The Lives of the Sophists*, William Heinemann, London/G.P. Putnam's Sons, New York, 1922.

Picknett, Lynn, *Mary Magdalene: Christianity's Hidden Goddess*, revised edition, Constable, London, 2004 (first edition 2003).

— *The Secret History of Lucifer: The Ancient Path to Knowledge and the Real Da Vinci Code*, Robinson, London, 2005.

Picknett, Lynn, and Clive Prince, *Turin Shroud: How Leonardo da Vinci Fooled History*, revised edition, Little, Brown, London, 2006 (*Turin Shroud – In Whose Image?*, Bloomsbury, London, 1994).

— *The Templar Revelation: Secret Guardians of the True Identity of Christ*, updated edition, Corgi, London, 2007 (Bantam Press, London, 1997).

— *The Masks of Christ: Behind the Lies and Cover-Ups About the Man Believed to be God*, Sphere, London, 2008.

Pico della Mirandola, (trans. Richard Hooker), *Oration on the Dignity of Man*, Washington State University website: www.wsu.edu:8001/~dee/REN/ORATION.

Plotinus (trans. Stephen MacKenna), *The Enneads*, abridged edition, Penguin, London 1991 (The Medici Society, London, 1917–30).

Plutarch, (trans. and ed. J. Gwyn Griffiths), *De Iside et Osiride*, University of Wales Press, Cardiff, 1970.

Polanyi, Michael, *The Tacit Dimension*, University of Chicago Press, Chicago, 2009 (Doubleday, New York, 1966).

Popper, Karl, *Unended Quest: An Intellectual Autobiography*, revised, Fontana, London, 1982 (originally published as 'Autobiography of Karl Popper' in Paul Arthur Schilpp, ed., *The Philosophy of Karl Popper*, Open Court, Illinois, 1974).

Purver, Margery, *The Royal Society: Concept and Creation*, Routledge & Kegan Paul, London, 1967.

Radin, Dean, *The Conscious Universe: The Scientific Truth of Psychic Phenomena*, HarperEdge, New York, 1997.

— 'Exploring Relationships Between Random Physical Events and Mass Human Attention: Asking for Whom the Bell Tolls', *Journal of Scientific Exploration*, vol. 16, no. 4, 2002.

— *Entangled Minds: Extrasensory Experience in a Quantum Reality*, Paraview, New York, 2006.

Ray, J.D., *The Archive of Hor*, Egypt Exploration Society, London, 1976.

Rees, Martin, *Just Six Numbers: The Deep Forces that Shape the Universe*, Weidenfeld and Nicolson, London, 1999.

Ridley, Matt, *Evolution*, revised edition, Blackwell Science, Malden, 2004 (first edition 1998).

Rose, Michael R., Harpid B. Passananti and Margarida Matos (eds.), *Methuselah Flies: A Case Study in the Evolution of Aging*, World Scientific Publishing Co., Singapore, 2004.

Rosencreutz, Christian, (trans. Joscelyn Godwin), *The Chemical Wedding of Christian Rosencreutz*, Phanes Press, Grand Rapids, 1991.

Rossi, Paolo, *Francis Bacon: From Magic to Science*, Routledge and Kegan Paul, London, 1968 (*Francesco Bacone: Della magia alla scienza*, Editori Laterza, Bari, 1957).

Rowland, Ingrid D., *Giordano Bruno: Philosopher/Heretic*, Farrar, Straus and Giroux, New York, 2008.

Schönborn, Christoph, 'Finding Design in Nature', *New York Times*, 7 July 2005.

Schueller, Gretel, 'Stuff of Life', *New Scientist*, no. 2151, 12 September 1998.

Shapiro, J. A., 'Bacteria are Small but not Stupid: Cognition, Natural Genetic Engineering and Socio-Bacteriology', *Studies in History and Philosophy of Biological and Biomedical Sciences*, 38(4), December 2007.

Singer, Dorothea Waley, 'The Cosmology of Giordano Bruno (1548–1600)', *Isis*, vol. xxxiii, pt. 2, no. 88, June 1941.

— *Giordano Bruno: His Life and Thought*, Henry Shuman, New York, 1950.

Smith, John Maynard, *The Evolution of Sex*, Cambridge University Press, Cambridge, 1978.

— *Did Darwin Get It Right? Essays on Games, Sex and Evolution*, Penguin, London, 1993 (Chapman & Hall, London, 1989).

Smolin, Lee, *The Life of the Cosmos*, Oxford University Press, Oxford, 1997.

— *The Trouble With Physics: The Rise of String Theory, the Fall of Science and What Comes Next*, Penguin, London, 2008 (Houghton & Mifflin, Boston, 2006).

Stockwood, Mervyn, (ed.), *Religion and the Scientists*, S.C.M. Press, London, 1959.

Susskind, Leonard, 'The Anthropic Landscape of String Theory', Cornell University, website (Feb 2003): arxiv.org/abs/hep-th/0302219.

— 'A Universe Like No Other', *New Scientist*, vol. 180, no. 2419, 1 November 2003.

— *Cosmic Landscape: String Theory and the Illusion of Intelligent Design*, Little, Brown, New York, 2005.

Teich, Mikaláš, and Robert Young, (eds.), *Changing Perspectives in the History of Science: Essays in Honour of Joseph Needham*, Heinemann, London, 1973.

Teilhard de Chardin, Pierre, *The Phenomenon of Man*, Collins, London, 1959 (*Le phénomène humain*, Éditions du Seuil, Paris, 1956).

Tompkins, Peter, *The Magic of Obelisks*, Harper & Row, New York, 1981.

Tuveson, Ernest Lee, *The Avatars of Thrice Great Hermes*, Bucknell University Press, London, 1982.

Usher, Peter, 'Shakespeare's Support for the New Astronomy', *The Oxfordian*, vol. 5, 2002.

van den Broek, Roelof, and Wouter J. Hanegraff, *Gnosis and Hermeticism from Antiquity to Modern Times*, State University of New York Press, Albany, 1998.

Walker, D. P., *Spiritual and Demonic Magic from Ficino to Campanella*, Warburg Institute, London, 1958.

— *The Ancient Theology: Studies in Christian Platonism from the Fifteenth to the Eighteenth Century*, Duckworth, London, 1972.

Watson, J. D., and F. H. C. Crick, 'Molecular Structure of Nucleic Acids: A Structure for Deoxyribose Nucleic Acid', *Nature*, vol. 171, no. 4356, 25 April 1953.

Weinberg, Steven, *The First Three Minutes: A Modern View of the Origin of the Universe*, Scientific Book Club, London, 1978 (André Deutsch, London, 1977).

— *Dreams of a Final Theory*, Hutchinson Radius, London, 1993.

Westfall, Richard S., *Never At Rest: A Biography of Isaac Newton*, Cambridge University Press, Cambridge, 1980.

Westman, Robert S., and J.E. McGuire, *Hermeticism and the Scientific Revolution: Papers Read at a Clark Library Seminar, March 9, 1974*, William Andrews Clark Library, Los Angeles, 1977.

Wheeler, John Archibald, with Kenneth Ford, *Geons, Black Holes, and*

Quantum Foam: A Life in Physics, W.W. Norton & Co., New York, 1998.

Wheeler, John Archibald, and Wojciech Hubert Zurek, *Quantum Theory and Measurement*, Princeton University Press, Princeton, 1983.

White, Michael, *Isaac Newton: The Last Sorcerer*, Fourth Estate, London, 1997.

Wickramasinghe, Chandra, *The Cosmic Laboratory*, University College, Cardiff, 1975.

—*A Journey with Fred Hoyle: The Search for Cosmic Life*, World Scientific Publishing Co., Singapore, 2005.

Williams, George C., *Sex and Evolution*, Princeton University Press, Princeton, 1975.

Yamauchi, Edwin M., *Pre-Christian Gnosticism: A Summary of the Proposed Evidences*, Tyndale Press, London, 1973.

Yates, Frances A., *Giordano Bruno and the Hermetic Tradition*, University of Chicago Press, Chicago, 1991 (Routledge & Kegan Paul, London, 1964).

—*The Art of Memory*, Ark, London, 1984 (Routledge & Kegan Paul, London, 1966).

—*The Rosicrucian Enlightenment*, Routledge & Kegan Paul, London, 1972.

—*The Occult Philosophy in the Elizabethan Age*, Routledge & Kegan Paul, London, 1979.

Yeffeth, Glenn, (ed.), *Taking the Red Pill: Science, Philosophy and Religion in* The Matrix, Summersdale Publishers, Chichester, 2003.

Yourgrau, Wolfgang, and Allen D. Breck, (eds.), *Cosmology, History, and Theology*, Plenum Press, New York, 1977.

ACKNOWLEDGEMENTS

Jeffrey Simmons, our agent and friend, for his usual unstinting help and support.

At Constable & Robinson: Andreas Campomar, Krystyna Green, Eryl Humphrey Jones, Jo Stansall and our editor, Leo Hollis, for his most constructive input.

David Bell, a dearly loved friend, for his often wicked humour, insight and support over many years. Much missed.

Keith Prince, for his customary invaluable assistance with research and especially for the many fruitful discussions that helped shape this book.

For their help, support and friendship: Deborah and Yvan Cartwright; Heather Couper; Jenny Boll; Carina Fearnley; Andrew Gough; Stewart and Katia Ferris; Nigel Henbest; Sarah Litvinoff; Moira Hardcastle; Jane Lyle; Neil McDonald; Sally Morgan; Craig and Rachel Oakley; James Pawson-Clark; Graham Phillips; Vlad and Mariana Sauciuc; Nick Spall; Mick Staley; Sheila Taylor; Oreste Teodorescu; Paul Weston; Caroline Wise.

Brian P. Copenhaver and Cambridge University Press for their kind permission to quote from Hermetica.

As always, the staff of the British Library, London. And the doctors and nursing staff of St Mary's Hospital, Paddington, London, without whom one of the authors wouldn't have been able to finish this book – or indeed anything else!

INDEX